D0075967

The Reproductive System

Other titles in
Human Body Systems

The Circulatory System by Leslie Mertz

The Digestive System by Michael Windelspecht

The Endocrine System by Stephanie Watson and Kelli Miller

The Lymphatic System by Julie McDowell and Michael Windelspecht

The Muscular System by Amy Adams

The Nervous System and Sense Organs by Julie McDowell

The Respiratory System by David Petechuk

The Skeletal System by Evelyn Kelly

The Urinary System by Stephanie Watson

The Reproductive System

Kathryn H. Hollen

HUMAN BODY SYSTEMS
Michael Windelspecht, Series Editor

Greenwood Press
Westport, Connecticut • London

Library of Congress Cataloging-in-Publication Data

Hollen, Kathryn H.
The reproductive system / Kathryn H. Hollen.
p. cm.—(Human body systems)
Includes bibliographical references and index.
ISBN 0-313-32449-2 (alk. paper)
1. Gynecology. 2. Generative organs, Female. 3. Generative organs, Male. 4. Human reproduction. 5. Reproduction. I. Title. II. Human body systems.
RG101.H73 2004
618—dc22 2004043638

British Library Cataloguing in Publication Data is available.

Library of Congress Catalog Card Number: 2004043638
ISBN: 0-313-32449-2

First published in 2004

Greenwood Press, 88 Post Road West, Westport, CT 06881
An imprint of Greenwood Publishing Group, Inc.
www.greenwood.com

Printed in the United States of America

The paper used in this book complies with the Permanent Paper Standard issued by the National Information Standards Organization (Z39.48–1984).

10 9 8 7 6 5 4 3 2 1

Illustrations, unless otherwise credited, are by Sandy Windelspecht.

For my mother and father, who bought us books

Contents

Color photos follow p. 108.

Series Foreword

Human Body Systems is a ten-volume series that explores the physiology, history, and diseases of the major organ systems of humans. An organ system is defined as a group of organs that physiologically function together to conduct an activity for the body. In this series we identify ten major functions. These are listed in Table F.1, along with the name of the organ system responsible for the activity. It is sometimes difficult to specifically define an organ system, because many of our organs have dual functions. For example, the liver interacts with both circulatory and digestive systems, the hypothalamus acts as a junction between the nervous and endocrine systems, and the pancreas has both digestive and endocrine secretions. This complex interaction of organs and tissues in the human body is still not completely understood.

This series is unique in that it provides a one-stop reference source for anyone with an interest in the human body. Whereas other references frequently cover one aspect of human biology, from anatomy and physiology to the prevention of diseases, this series takes a more holistic approach. Each volume not only includes a physiological description of how the system works from the cellular level upward, but also a historical summary of how research on the system has changed since the time of the ancients. This is an important aspect of the series and one that is frequently overlooked in modern textbooks. In order to understand the successes and problems of modern medicine, it is first important to recognize not only the achievements of the past but also the misunderstandings and challenges of the pioneers in medical research.

For example, a visit to any major educational institution reveals large lec-

TABLE F.1. Organ Systems of the Human Body

Organ System	General Function	Examples
Circulatory	Movement of chemicals through the body	Heart
Digestive	Supply of nutrients to the body	Stomach, small intestine
Endocrine	Maintenance of internal environmental conditions	Thyroid
Lymphatic	Immune system, transport, return of fluids	Thymus
Muscular	Movement	Cardiac muscle, skeletal muscle
Nervous	Processing of incoming stimuli and coordinates activity	Brain, spinal cord
Reproductive	Production of offspring	Testes, ovaries
Respiratory	Gas exchange	Lungs
Skeletal	Support, storage of nutrients	Bones, ligaments
Urinary	Removal of waste products	Bladder, kidneys

ture halls, where science instructors present material to the students on the anatomy and physiology of the human body. Sometimes these classes include laboratory sessions, but in the study of human biology, especially for students who are not bound for professional schools in medicine, the student's exposure to human biology typically centers on a two-dimensional graphic. Most educators accept this process as a necessary evil of the educational system, but few recognize that, in fact, the large lecture classroom is the product of a change in Egyptian religious beliefs before the start of the current era. During the decline of the Egyptian empires and the simultaneous rise of the ancient Greek culture, the Egyptian religious organizations began to forbid the dissection of the human body. This had a twofold influence on medicine. First, the ending of human dissections meant that medical professionals required lectures from educators, instead of participation in laboratory-based education, which led to the birth of the lecture hall. The second consequence would plague modern medicine for a thousand years. Stripped of their access to human cadavers, researchers studied other "lesser" animals and extrapolated their findings to humans. The practices of the ancient Greeks were passed on over the ages and became the basis for the study of modern medicine. These traditions continue to this day throughout the educational institutions of the world.

The history of human biology parallels the development of modern science. In the seventeenth century, William Harvey's study of blood circulation challenged the long-standing belief of the ancient Greeks that blood was produced in the liver and consumed in the tissues of the body. Harvey's pioneering experimental work had a strong influence on others, and within a century the legacy of the ancient Greeks had collapsed. In the eighteenth century a group of chemists who focused on the chemical reactions of the human body, called the iatrochemists, began to apply chemical laws to human physiology. They were joined by the iatrophysicists, who believed that the human body must operate under the physical laws of the universe. This in turn led to the beginnings of organic chemistry and biochemistry in the nineteenth century, as scientists focused on identifying the building blocks of living cells and the chemical reactions that they utilize in their metabolism.

In the past century, especially in the last three decades, the rapid advances in technology and scientific discovery have tended to separate most sciences from the general public. Yet despite an ongoing trend to leave the majority of the physical sciences to the scientists, interest in the human biology has actually increased among the general population. This is primarily due to medical discoveries that increase not only lifespan but also healthspan, or the number of years that people live disease free. But another important aspect of this trend is the desire among the general public to be able to ask intelligent questions of their physicians and seek additional information on prescribed medications or procedures. In many cases this information serves as a system of checks and balances on the medical profession, ensuring that the patient is kept well informed and aware of the fundamentals regarding the procedure.

This is one of the most remarkable ages in the study of human biology. The recently announced completion of the Human Genome Project is an indication of how far biology has progressed. Barely fifty years ago, scientists were first discovering the structure of DNA. They now are in possession of an entire encyclopedia of human genetic information, and although they are not yet exactly sure what the content reveals, scarcely a week goes by without a researcher announcing a medical discovery that was made possible by the availability of the complete human genetic sequence. Coupled to this are the advances in the development of pharmaceuticals and treatments that were unheard of less than a decade ago.

But these benefits to society do not come without a cost. The terms stem cells, cloning, and gene therapy no longer belong to the realm of science fiction. They represent advances in the sciences that may hold the key to increased longevity. However, in many cases they also produce ethical and moral questions of society: Where do medical researchers obtain the embryonic stem cells for their work? Who will determine if humans can be cloned? What are the risks of transgenic organisms produced by gene therapy? These are just a few of the potential conflicts that face modern society. Only a well-educated general public can intelligently survey the pros and cons of an ethical or moral decision regarding medical science. Armed with information, concerned people can participate in the democratic process of informing their elected officials of their concerns. Science education is an important aspect of citizenship, and thus the need for series such as this to present information to the general public.

This volume covers the biology of the reproductive system, which is responsible for far more than the development of sexual characteristics. The hormones it produces influence the operation of practically every cell of the body, and the system itself influences the reproductive glands, the wiring of our brains, the deposition of adipose tissue on the body, and the way that we age. Of all human organ systems, it is the one most different between males and females. The female reproductive system transforms the female anatomy and physiology in preparation for childbirth and again after menopause. The male reproductive system is responsible for the production of millions of sex cells daily, throughout the life of the individual. In this volume, the author examines the various aspects of reproduction from the cellular to organismal level, and takes a long look at some of the more recent advances in the study of human reproduction. Readers of this work will find information on new discoveries in artificial fertilization techniques, as well as an in-depth look at what is now known about the sex hormones.

The ten volumes of *Human Body Systems* are written by professional authors who specialize in the presentation of complex scientific ideas to the general public. Although any book on the human body must include the terminology and jargon of the profession, the authors of this series keep it to a minimum and strive to explain the concepts clearly and concisely. The series is ideal for the public libraries, as well as for secondary school and introductory college libraries. In addition, medical professionals or anyone with an interest in human biology would find this series a useful addition to their personal library.

Michael Windelspecht
Blowing Rock, North Carolina

Acknowledgments

I would like to thank Dr. Michael Windelspecht for the opportunity to write this volume and for his generous, balanced counsel. I applaud Greenwood Press for committing resources to this educational series and am privileged to contribute to it. It is always a pleasure to write about science, but it is all the more rewarding to write for students.

Family members who lent precious time to reviewing portions of the manuscript include Jean Howell, Martha Howell, Kadi Row, and Amy Row Lenert. I am grateful for their help and have tried my best to incorporate their thoughtful suggestions. Jeanne Kelly and Wendy Kirwan deserve special thanks for their sustained support and assistance with eleventh-hour assembly of the final manuscript, and I appreciate some helpful comments made by Janet Manthos and Maggie Clarridge. During the development of the black and white illustrations, I discovered the pleasure of working with Sandy Windelspecht, who produced them with such professionalism.

It is for Brian, my husband and business partner, that I reserve my deepest appreciation, not for the hours he spent reading drafts or drawing and redrawing color illustrations, but for his unwavering enthusiasm and dependably cheerful support, at home and at work. For these essential contributions, this book is also his.

Introduction

The systems that animate the human body do not operate independently of one another. They are united by a network of blood, nerves, and tissues that give them substance and function, and, in this respect, the reproductive system is no different from its counterparts. But in another respect, it is unique, for it is the only system in the body dedicated exclusively to the continuation of the species. This single fact dictates a dual direction for the material in this volume. On the one hand, it must address the biology of the reproductive system that keeps the human race going. On the other, it must also consider, albeit much more briefly, how tapping into human DNA can so transform genetic identity that the definition of what it means to be human may unravel.

The first duty of this book, then, is to present clear, factual, and comprehensive information about the function of the reproductive system. The description of the tissues and roles of different organs, the development of a fetus, sexual maturation in adolescents, and the impact of aging meet this obligation in the first four chapters. Because the information presented there will likely raise many questions about how today's wealth of knowledge was accumulated, the next four chapters look at human reproductive biology from an historical perspective, a narrative that weaves a timeline of discovery. The remainder of the chapters discuss diseases and other disorders that impede or shut down proper functioning of a system that is often taken for granted until it fails, as it sometimes does when couples find themselves unable to conceive. Chapter 10 looks at infertility, offering a detailed description of how assisted reproductive technologies (ART) help millions have children when all other attempts to conceive have failed.

Introducing genes, the molecules within the cellular nuclei that write the script defining human as human, was the logical starting point for Chapter 1, which describes how coiled DNA from a male and a female parent combine to produce a new organism genetically different from any human being ever created (including its identical twin). The breathtaking choreography of this recombination is called *meiosis*, and it has been happening in millions of species, millions of times a day, for eons. Meiotic division, with its exchange of genes, is the defining feature of sexual reproduction, and helping readers develop a clear understanding of how it works is the reason it is featured in color at the center of the book.

Succeeding chapters describe the series of steps by which male and female reproductive organs carry out their essential roles, including the complex physiology of attraction, desire, and sexual fulfillment. Drawings accompany the text to illustrate how an embryo, once conceived, develops from a collection of microscopic cells into a full-term baby. But the cycle of reproduction initiated by genetic recombination does not end with birth, because a baby will, in all likelihood, reproduce someday. The body's machinery that makes this possible is already present in a newborn, but it is dormant, emerging during puberty when the child reaches sexual maturity. Described in Chapter 4, this maturation—the ability to conceive another human being—marks the completion of a reproductive cycle that begins in humans at the moment of their own conception.

In examining the steps and missteps that have taught modern scientists what they know about the reproductive system and what it means, in biological terms, to be human, the next section explores some of history's most profound insights and developments. Chapters 5 and 6 show that, as scientific investigation across the centuries came together in the mid-to-late-1800s, the lines dividing genetics, embryology, and evolution began to blur. Elucidating the structure of DNA in the 1950s solidified much of the previous century's theory and launched a biotechnical revolution that represents the threshold of what truly may be a brave new world. With scientists already capable of inserting artificial chromosomes into mice to alter their genome (and thus their essential "mouseness"), Chapter 7 raises some questions about what such technology could mean for human beings.

The very genes that are the agents of heredity are also the agents of the body's future design and function. Whether—and how—scientists, politicians, ethicists, physicians, and others tap the potential that genes offer for good or harm has enormous consequences for the human race. This required a judgment to be made about how deeply to examine evolution and futuristic technology in these chapters. On the one hand, neither genetic origins nor posthumanism can be said to relate directly to reproductive biology in terms of how cells and organs work together to produce offspring. On the other, DNA cannot be regarded as just another molecule comprising the larger group of molecules that is the human body. It is the thread that links all of humanity, carrying human history and—unless it is diluted into oblivion by the repeated introduction of nonhuman genes—the promise of its future. Artificially reengineering the molecules exchanged during sexual reproduction changes the science of reproduction and, in a larger sense, its product. This provided the rationale for introducing the cultural implications of coexistence with "cyborgs," quasi-human organisms partially manufactured with artificial components. Not only must humans understand where they themselves came from, they must also understand where posthuman technology may lead them—and how. Discussions of human beginnings and, especially, futuristic scenarios became relevant within this framework. Current controversies surrounding embryonic stem cell research and germline engineering highlight both the threat and the promise of exploiting reproductive biology.

It was appropriate to include in Chapter 8 an historical overview of medical advances, from prehistory to modern times, in treating diseases of the reproductive tract. As the chapter's accompanying chronology shows, very little surgery was attempted on the body's internal organs until well into the eighteenth century, when dissection of human cadavers taught barber-surgeons enough about anatomy for them to attempt invasive surgery on patients. The chapter serves as a prelude to the next section, which studies the most common disorders affecting the reproductive system today and the most widely applied therapies used in treatment. Among the conditions discussed are, of course, cancer and sexually transmitted diseases, along with some of their causes. The chapter presents a general overview of medical problems and therapeutic approaches, but is not intended to serve as a guide to the diagnosis or treatment of disease. Such advice should always be obtained from a physician or other appropriate healthcare professional.

At the end of the volume is a list of commonly used acronyms followed by a glossary of unfamiliar terms. To minimize the disruptive need to flip back and forth, explanations of new terms were provided within the text whenever possible. A list of organizations and Web sites and a bibliography list sources that supplement this volume or student coursework.

Just as it has transformed understanding of human genesis and design, reproductive science is itself being transformed, opening rich, thrilling avenues of exploration that today's biology students will someday pursue. Assisted by unimaginably sophisticated tools and technology, they will gain increasing control over reproductive function, but it will likely be a very long time before technological innovations will surpass—in design or efficiency—the remarkable system by which humans have been reproducing for tens of thousands of years.

INTERESTING FACTS

- The proportions of a developing fetus change dramatically as it grows. The size of its head at the ninth week of development is almost one-half the total length of its body; an adult's, in comparison, is only about one-eighth the total length of its body.

- Increased age in women increases their likelihood of having twins because the zona pellucida tends to be harder in older women and causes the inner cell mass to break into clumps as it "hatches."

- About 50 percent of conceptions fail between fertilization and implantation due to abnormalities in the specialized cells required for implantation.

- A fetus can detect light coming from outside the mother's body; it will turn to follow a light moving outside and across the abdomen.

- The first successful vaginal hysterectomy for the cure of uterine prolapse was self-performed by a peasant woman in the seventeenth century. She slashed off the prolapse with a sharp knife, surviving the hemorrhage to live out the rest of her life.

- The brain of a child malnourished in the womb and in infancy may be 60 percent smaller than that of a normal child.

- Babies in the womb dream, which contributes to brain development.

- Several theories surround the discrepancy in size between the sperm and the egg. One is that the egg is larger because it contributes most of the bulk to the embryo while the small sperm must be an agile, mobile hunter of the ovum. Another suggests that sperm have to be tiny to be produced in quantity; the more produced, the greater the likelihood that only superior sperm will reach the egg.

- Genes from a male parent are required for placental development in humans. Therefore, even if an unfertilized egg

divides and grows, it cannot develop a lifeline to the mother and will die. In a unique (so far as is known) case in Scotland, a young boy has been discovered to be a "male/parthenogenic chimera"—that is, half of his genetic makeup is that of a normal male and the other half is parthenogenic. Scientists have deduced that his mother's egg began to divide just before fertilization into two cells. One cell was then fertilized with sperm, creating a normal genomic blueprint from both parents. The unfertilized portion continued to develop but all the descendant cells contained only genetic input from the mother. The father's genes allowed a placenta to develop, but the child is a chimera, an organism that carries cell populations having different genetic makeup. He has some physical abnormalities and mild learning difficulties, reinforcing the widespread belief that genetic imprinting is required for normal development.

- In 1512, Frenchman Arnaud de Villeneuve claimed that if a garlic clove were inserted into a woman's vagina and garlic could then be smelled on her breath, her reproductive system was functioning normally and she could become pregnant.

1

Sex and Reproduction

Human beings have sex to reproduce, but there's another reason, one that's critical to the vitality of the species—and to that of other species as well. In fact, sex is a fundamental feature of the reproductive life of most animals, although some of them engage in it differently from humans. The frog, for example, lays her eggs near a river or stream for later fertilization by any compatible male frog that happens by, while humans customarily have reproductive sex together, at the same time. People are very different from frogs in another way, too, in that human sex can result in serious consequences like unwanted pregnancy, disease, social disgrace, moral or religious disapproval, or legal difficulties. **Asexual reproduction** or cloning (see Chapter 7) would likely be easier; hydra and sponges are able to grow buds genetically identical to themselves with far fewer consequences. Flatworms and sea stars reproduce by generating new organisms from their own detached parts. Why do people complicate their lives with sex and all that it entails?

The answer is simple: genetic diversity, the mingling of genes from two different parents to create genetically unique offspring. This is the very definition of sexual reproduction, a means of propagation that, unlike the asexual method, reduces the chance that harmful mutations will be passed on because the normal gene from one parent can dominate the defective gene from the other parent. Since not all mutations are bad, however, sexual reproduction is also an avenue for passing on beneficial traits, such as genetic adaptations that might, for example, improve one's resistance to environmental **pathogens**. And just in case the species neglects to engage in the reproductive activity that biology has so thoughtfully designed for it, renewed motivation arises from the persistent nudgings of the **libido**—that compelling drive that, some would argue, is second only to the need for food and water in the urgency of its demand.

In a strictly biological sense, sex is fundamental to reproduction in humans, as in many other species, because the offspring gain genetic resistance to disease and threatening mutations are minimized or purged. Mutations are widely discussed in Chapters 5–8 of this volume, which explore their significance to evolution.

Human sexuality is the manifestation of an ancient biological imperative programmed into the genes that triggers desire, orchestrates the physiological changes the human body undergoes to produce sperm or sustain pregnancy, and fosters nurturing instincts that prompt caring for and protecting offspring. It is intrinsic to a person's existence and its influence is so pervasive in human interactions that, throughout recorded history and with greater or lesser success, societies have legislated it, politicized it, demonized it, sanctified it, sanitized it, or suppressed it. Oblivious to cultural constraints, however, biology has prevailed and humans continue to multiply despite wars, famines, and epidemics; despite the pain and danger of childbirth, the heartache and sacrifice that can accompany parenthood, or the economic hardship of raising children; and despite alarming predictions of how overpopulation will further burden this already endangered planet. It is clear that the libidinous urges emanating from the primitive self are very powerful indeed.

Uncovering the source of these instincts has led scientists deep into the cell where they are studying how molecular activity drives human reproduction and, for that matter, other human behavior. They are finding the laboratory of the cell so rich and its biochemical communications so complex that new disciplines in the life sciences such as proteomics, the study of the structure and function of proteins, have arisen out of the need to address the profound questions their investigations raise. It is already known that well before a woman begins to experience symptoms of pregnancy, most of her child's heritable characteristics have been determined. The color of its eyes and hair, the contours of its face and the stature of its body, even aspects of its temperament and personality are traits that were laid down at the moment of conception. As a fertilized ovum divides until it is a bundle of cells that implants itself in the lining of the womb, the genetic instructions for the organism to develop, to mature, and to transmit its own genes to its progeny when biology deems the time is right has been programmed into each of those cells. It is only in recent years that science has begun to unravel the complexities of how genes and cells communicate and interact.

GENES

Genes, the human **genome**, genetic engineering, sequencing the genome—most people have heard and read a great deal about these subjects in recent years. Genes determine the characteristics that define the species as human: creatures who walk erect, who are born with two legs and arms and hands, and who are capable of learning to read and write. Genes also determine if a person's legs will be short or long, his nose large or small, his hair blond or black; each of these traits makes every person unique even though 99.999 percent of nearly all humans' genes are same. (Identical siblings like twins, however, start out in life with 100 percent of the same genes.) When deoxyribonucleic acid (DNA) replicates and accidentally drops a critical molecule from the gene during transcription (see "RNA, Transcription, Translation, and the Genetic Code"), or when environmental assaults such as exposure to cigarette smoke accumulate in the cells' DNA, the damage can be serious enough that, later in life, the affected twin might develop a genetically related disease while the other does not.

Although it is generally understood that every person is the product of unique combinations of parental genes, not everyone understands the intricate details: what genes are made of, where they reside in the cell, and how cell division ensures genetic diversity in the species.

Like all living things, the human body is made up of chemicals, tiny molecules such as the four **amino acids** comprising genes. Called bases, these amino acids (adenine, cytosine, guanine, thymine) reside in the cell as part of the coiling DNA that forms diffuse nuclear matter known as **chromatin**. Two bases, adenine and thymine (A and T), always form one pair, and cytosine and guanine (C and G) always form the other. These are **complementary base pairs** that, held loosely together by hydrogen bonds, construct the "rungs" of DNA's familiar ladder or spiraling double helix (see Figure 1.1). When a cell begins to divide, DNA twists and condenses out of chromatin to organize into chromosomes and to replicate (see color illustrations at the center of the book). This is how the cell's copy will receive the same DNA—that is, the same chromosomes and genes. In turn, as these **daughter cells** divide, they produce exact replicas of themselves. Thus all the genetic information of the first cell is transferred to every descendant.

Although the word "gene" is used as a singular noun, Figure 1.1 illustrates that a gene actually represents the components of a given piece of DNA that act as a single unit to direct an activity. The genes utilize twenty different amino acids that, in turn, can construct about 250,000 different proteins. This is known as encoding; that is, the gene encodes for a protein, an enzyme perhaps, which in turn unleashes a cascade of chemical reactions triggering cellular activity. The gene that initiated this activity is said to have been **expressed**, that it has been "switched on." Recent research indicates that proteins called **histones** on the DNA tell genes when to turn on or off. In **embryogenesis**, the early stages of human development in the womb, they are switched on to tell the cells what kind of tissue to become, transforming a cell into a muscle cell, for instance, and instructing the new

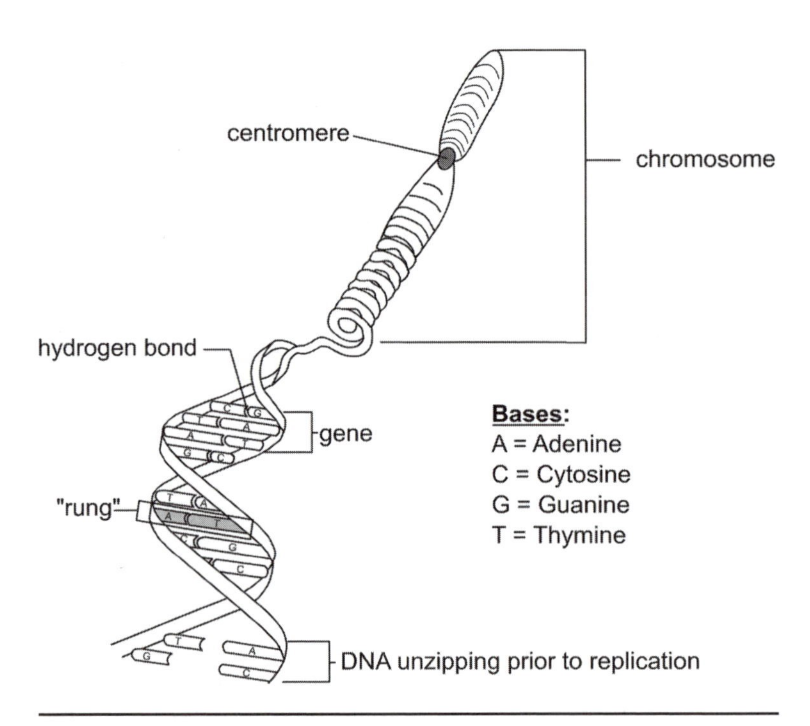

Figure 1.1. The DNA double helix uncoiling from a chromosome.
Note that each base is paired with its complementary base. When DNA replicates, it "unzips" between the paired bases and unites with a spiraling complement of itself, which has been synthesized in the cell by a complex series of steps carried out by RNA.

cell how to do its job. Every muscle cell division thereafter will produce a muscle cell, but it was a series of genes that told the *first* cell what to become. Cells in muscle tissue look different from those in **epithelial** tissue, just as these two look very different from a **neuron**, a type of brain cell (see Figure 1.2). How the cells are instructed to become different kinds of cells and thus to take on different functions is a subject of intense interest, not just within the research community but among the lay public, politicians, bioethicists, and the medical establishment. Because that first cell, called a *stem cell*, has the ability to differentiate into any other kind of cell the genes so order, its therapeutic promise is tantalizing. Researchers are developing techniques to cultivate the cells in the laboratory and to manipulate their differentiation with the goal of growing human organs—**in vivo** or **in vitro**—to replace diseased ones. But before therapeutic breakthroughs of this kind are likely to be commonplace, a significant ethical hurdle must be addressed: the stem cells with the greatest potential for differentiation can be obtained only from human embryos. Although adult stem cells can be harvested from certain portions of the body, most scientists find them of limited value. (See Chapter 7 for a fuller discussion of both the promise and the ethics of embryonic stem cell research.)

Sequencing the human genome simply means mapping all the genes—each specific group of bases that, once switched on, directs the manufacture of the proteins that drive cellular functions—in human DNA to determine which group of letters represents which gene. While it is tempting to assign one function to one gene, it is usually not that simple, although there are instances, such as in **Huntington's chorea** or **sickle cell anemia**, in which one causative gene can be identified. Usually, however, it appears that different genes may interact with external features in the environment or may combine to predispose someone to a disease or, conversely, to confer resistance to it. Nevertheless, links can be made between the influence of specific genes on specific characteristics or diseases, and some of these will be discussed later in this book.

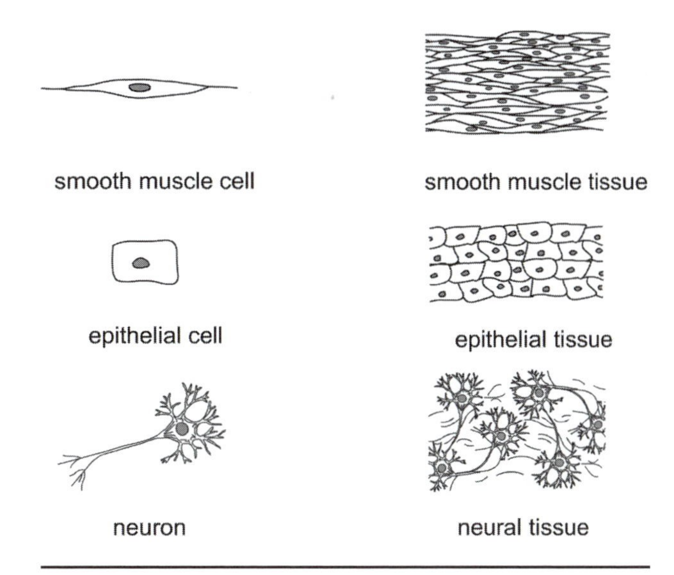

Figure 1.2. Samples of different kinds of cells and tissues.
Epithelial cells give rise to epithelial tissue, which forms the inside and outside surfaces of many organs; the muscle cell depicted here produces smooth muscle tissue, which comprises involuntary muscles like those of the intestines; and neurons (nerve cells) like these feature dendrites, branching arms reaching out from the body of the neuron to receive impulses from the nervous system.

Interestingly, although the human genome contains roughly three billion base pairs, scientists disagree about how many of these are genes. The U.S. Human Genome Project, jointly launched in 1990 by the U.S. Department of Energy and the National Institutes of Health to identify all the genes in human DNA, announced in 2003 that there are more than 30,000. Most scientists were surprised by the low number and by the discovery that people share so many of their genes with animals, especially chimpanzees (an intriguing finding referred to in the discussion of evolution in Chapter 6). It is this paucity of genes regulating something as complicated as the human body that has led many experts to believe that initiating cellular activity requires the cooperation of several genes.

That leaves a great deal of excess DNA, called introns or "junk DNA," whose role puzzles scientists. Many believe that these non-coding sequences, probably enough in each person's 100 trillion cells to reach the sun and back, are made up of meaningless pieces of viruses and other genetic debris whose sole value lies in providing an evolutionary fossil record of human ancestry. Others maintain that they may represent crucial coding sequences that control other genes in as-yet-undetermined ways and thus multiply the effect of all.

CHROMOSOMES

Figure 1.1 depicts the DNA "ladder," showing the two spiraling "uprights" made up of **phosphates**, sugars, and bases. One phosphate, sugar, and base form a nucleotide (see Figure 1.3) that matches up to its complement, another group of three molecules; if one piece of the DNA is made of phosphate, sugar, and adenine (one of the four bases), its counterpart would be attached at the rung to thymine, sugar, and phosphate. In every cell (except mature red blood cells, which discard their nuclei during human development to carry more oxygen throughout the blood), about six feet (180 centimeters) of DNA resolve out of chromatin into forty-six coiled chromosomes that form twenty-three pairs. Each pair is comprised of two **homologous** chromosomes—that is, they carry essentially the same genes, but one chromosome is from the mother and one chromosome is from the father. An example is the gene for eye color. Sup-

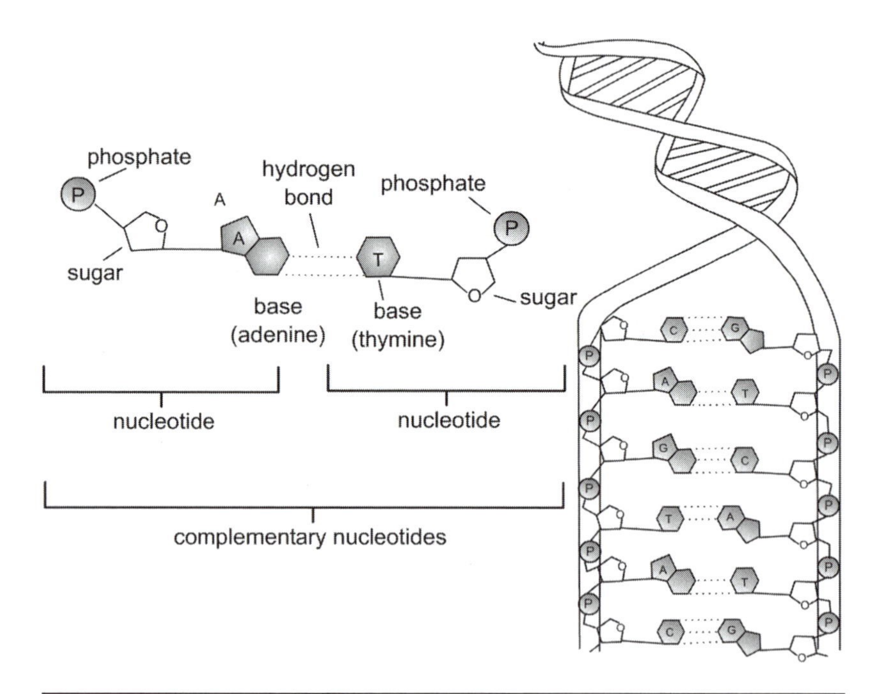

Figure 1.3. Nucleotides in DNA.
A phosphate, a sugar, and a base make up each nucleotide. They form a part of DNA that will pair up with its complement during DNA replication. Note in the representation of the double helix that there are two dotted lines between adenine and thymine to denote the two hydrogen bonding sites that link them, and three dotted lines between cytosine and guanine to indicate that they have three bonding sites. Note too that adenine and guanine have double-ring chemical structures, while cytosine and thymine have single-ring structures.

pose a person inherited a gene for blue eyes from her mother and a gene for brown eyes from her father. The genes for eye color in this case are called **alleles** because, while they're essentially the same gene—they both transmit eye color characteristics—one encodes for one color on the mother's chromosome and the other for a different color on the father's. Thus alleles are different expressions of the same gene on homologous chromosomes. Chapter 5 discusses dominant and recessive genes to explain which eye color the offspring inherits.

So twenty-three chromosomes from one parent pair up with their homologs from the other parent. Each chromosome *pair*, however, is different from every other pair. Twenty-two of these pairs are called autosomes; the twenty-third pair comprises the sex chromosomes, so called because they determine the sex of the offspring and carry the genes for **sex-linked inherited characteristics** like color-blindness. In women, the sex chromosome is called an X chromosome; in men, it's a Y chromosome. Figure 1.4 is a

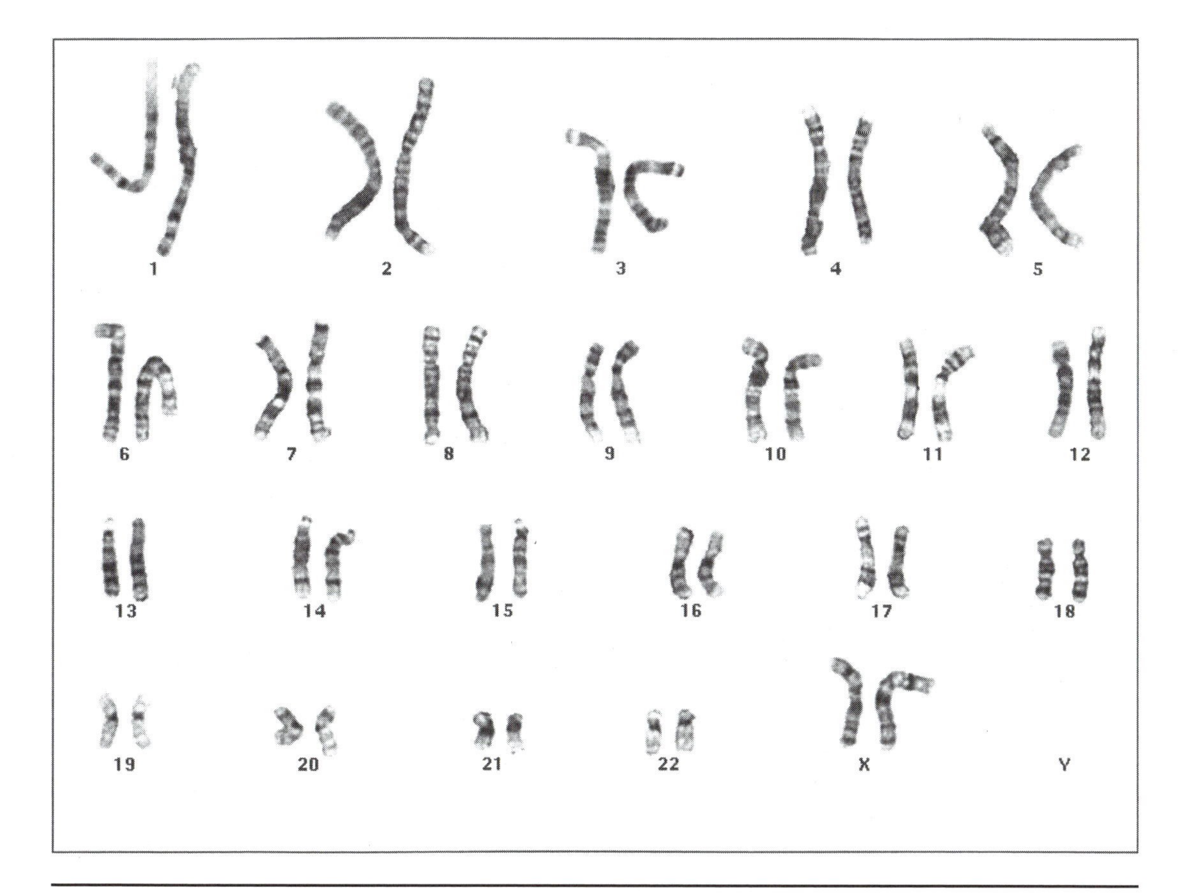

Figure 1.4. A karyotype.
A microphotograph of the actual chromosomes of a human (female, in this case) arranged in descending order of size. © Kathryn Hollen

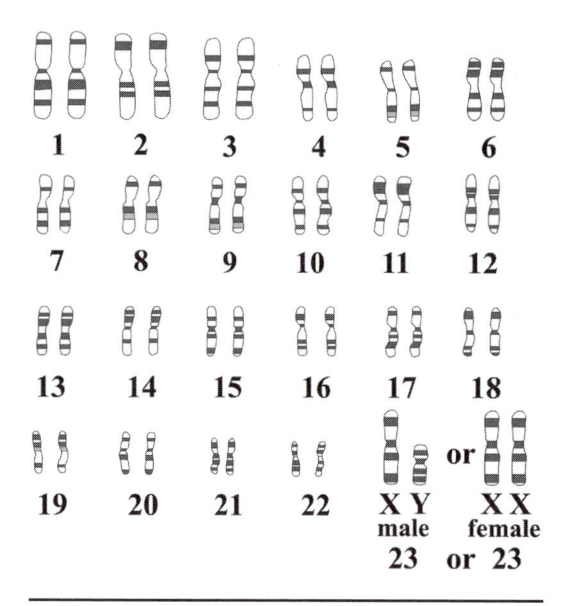

Figure 1.5. A graphic representation of the forty-six human chromosomes that shows banding patterns.

karyotype of all twenty-three pairs of chromosomes inside the cell of a human female (i.e., with two X chromosomes) and Figure 1.5 is a graphic representation of the chromosomes inside the cell of either a male or a female (with either XY or XX chromosomes, respectively). Notice the difference in size between the X and Y chromosomes; this discrepancy occurs because the Y chromosome has no Y mate in the cell (as the X chromosome does) from which to obtain new copies of damaged or discarded genes. Over time, it had been believed, the Y chromosome was "rotting" away. New evidence has shown, however, that it curls back onto itself to, in effect, copy itself and create a backup of critical genes.

Also note the banding patterns elicited by staining or dyeing techniques that reveal different areas and help identify the location of the **centromere**, or "waist," on each. Featured prominently in cell division, the centromere distinguishes the shorter and longer sections of the chromosomes that are called, not surprisingly, the short arm and the long arm. Banding allows researchers to pinpoint similarities in gene patterns and thereby establish biomarkers, particular patterns that might relate to a genetic sequence of concern, such as predisposition to disease, or to something benign, such as hair color. The look-alike banding patterns on paired chromosomes illustrate why they are homologous.

CELL DIVISION: MEIOSIS AND MITOSIS

Humans have two kinds of cells: body (somatic) cells, which form all of their tissues, and sex (germ) cells, which in women are eggs and in men are sperm. Although a human being's body cells are building and repairing tissues like heart muscle or bones throughout that person's life, sex cells have prepared themselves for one thing only—reproduction. The decision about which cells became body cells and which became sex cells was a differentiation decision made early in embryogenesis. And while it is true that all but the red blood cells contain exactly the same forty-six chromosomes, there comes a time when the sex cells must decrease their number to twenty-three, or a haploid (n) number, so that, once male and female sex cells unite in fertilization, the original diploid number of chromosomes, forty-six (or 2n), will be restored in the new organism.

Sex cells divest themselves of half of their chromosomes, and chromosomes mix up their genes to impart diversity to new cells, during a special type of cell division called *meiosis*, or meiotic division. Like regular cell division, meiosis relies partially on ribonucleic acid (RNA), a chemical closely related to DNA and, some suggest, its evolutionary precursor. In humans, RNA carries out essential missions such as relaying messages from the genes to the cells and replicating DNA so there is a copy available for the new cell (see "RNA, Transcription, Translation, and the Genetic Code").

Meiosis and mitosis are depicted in color illustrations in the center of this book; remember that the function of meiosis is to mix both parents' genes and to reduce the number of chromosomes in the resulting cells by half, whereas, in mitosis (mitotic division), the full complement of chromosomes from the parent cell is passed on to the daughter cell *with no change in genetic material.* Thus meiosis ensures the progeny will receive a diverse set of genes and, at the end of meiotic division, the germ cells will have be-

RNA, Transcription, Translation, and the Genetic Code

Having a slightly different chemical structure from DNA, which stores genetic instructions, RNA interacts with **ribosomes**, protein-making machinery in the cell, to **synthesize** proteins. When the genes order a particular protein to be made, the portion of DNA containing the amino acids needed to build it is **transcribed** into "messenger RNA" (mRNA) that tells the cell which protein to manufacture. "Transfer RNA" (tRNA) then brings the mRNA to the ribosome, where it is **translated** into the amino acid sequence that "ribosomal RNA" (rRNA) uses to build the protein.

Three of four nucleotide bases (A, C, G, T) make up a **codon**, which in turn comprises each amino acid (see Figure 1.6). Some amino acids have more than one codon; as the chart shows, AGA or AGG or CGT or CGC or CGA or CGG can specify arginine. Other amino acids have only one. Since codons are frequently repeated along a strand of DNA, there are "start" and "stop" codes that signal rRNA to stop adding amino acid chains to the protein being synthesized.

Only sixty-four combinations of bases can be created to specify twenty essential amino acids. As the chart shows, there is some room for error in coding since most codons can be changed by a single letter or two and still specify the same amino acid. However, if the *wrong* base in the codon is changed or dropped, the "misspelling" can result in a mutation.

	T	C	A	G
T	TTT Phenylalanine TTC Phenylalanine TTA Leucine TTG Leucine	TCT Serine TCC Serine TCA Serine TCG Serine	TAT Tyrosine TAC Tyrosine TAA STOP CODE TAG STOP CODE	TGT Cysteine TGC Cysteine TGA STOP CODE TGG Tryptophan
C	CTT Leucine CTC Leucine CTA Leucine CTG Leucine	CCT Proline CCC Proline CCA Proline CCG Proline	CAT Histidine CAC Histidine CAA Glutamine CAG Glutamine	CGT Arginine CGC Arginine CGA Arginine CGG Arginine
A	ATT Isoleucine ATC Isoleucine ATA Isoleucine ATG Methionine START	ACT Threonine ACC Threonine ACA Threonine ACG Threonine	AAT Asparagine AAC Asparagine AAA Lysine AAG Lysine	AGT Serine AGC Serine AGA Arginine AGG Arginine
G	GTT Valine GTC Valine GTA Valine GTG Valine	GCT Alanine GCC Alanine GCA Alanine GCG Alanine	GAT Aspartic Acid GAC Aspartic Acid GAA Glutamic Acid GAG Glutamic Acid	GGT Glycine GGC Glycine GGA Glycine GGG Glycine

Figure 1.6. The genetic code.
The above three-letter nucleotides are the codons of DNA showing the amino acids for which they code. mRNA converts thymine (T) into uracil (U), so when the genetic code is spelled out in RNA, each "T" becomes "U"; for example, GTT becomes GUU.

come **gametes**, reproductive cells containing only twenty-three chromosomes. As telophase II depicted in the color insert shows, the four haploid cells resulting from meiosis are genetically different from each other. When meiosis is completed, the **zygote** will inherit traits from both mother and father because the nucleus of its father's sperm contained twenty-three chromosomes that were comprised of genes the father received from *his* mother and father, and the nucleus of its mother's egg also contained, at fertilization, twenty-three chromosomes that carry genes she received from *her* mother and father. In this way, the zygote received forty-six chromosomes worth of its parents' (and *their* parents', and so on) genetic material. Someday, as a sexually mature adult, what is now the zygote may contribute twenty-three of its chromosomes to its own child.

In the male, there's always a sufficient supply of sperm maturing in the testes, because meiotic division has already been completed and four haploid sperm cells from each germ cell reside there. In the female, however, a specific number of egg cells are produced only by mitotic division early in the development of the female embryo, and it isn't until the maturing female reaches puberty that, at ovulation, one egg a month will complete its first meiotic division and discharge excess chromosomes not into a second egg but into a polar body, a poor cousin to the ovum. Meiosis is arrested at that point and it won't be until fertilization that the ovum completes its second meiotic division, retaining most of the original cellular **cytoplasm** for itself while the extra chromosomes and remaining cellular material are consigned to another polar body or two, all of which are nonfunctional and will simply degenerate. Immediately after fertilization, the **pronuclei** of the sperm and egg cells merge and become the nucleus, containing forty-six chromosomes, of the newly conceived organism, the zygote.

SUMMARY

Within the molecular makeup of the human cell lie the genetic instructions that drive the machinery of the human body. Ensuring the vitality of the species helps ensures its survival, so nature has engineered it to reproduce by sexual means. Reducing diploid cells to haploid by meiosis sets the stage in which haploid gametes fuse their pronuclei to create a genetically unique organism with the full complement of chromosomes.

2

The Reproductive Organs

The last chapter briefly discussed differentiation, how certain cells become nerve cells, others become muscle cells, and still others become sex cells. In the early embryo, the latter are called *primordial* germ cells. They migrate to an undifferentiated area of tissue called the genital ridge where they multiply, becoming **oogonia** in the female embryo's developing ovaries or **spermatogonia** in the male's developing testes, or testicles. The essence of human life, they are poised to complete their development whenever the body summons them.

FEMALE REPRODUCTIVE ORGANS

Although several million primordial germ cells are produced during the embryo's development in the womb, hundreds of thousands die en route to the ovary and afterwards, during a girl's growth. This is due to apoptosis, or cell suicide, the body's normal biological response to excessive cell proliferation. A female born with about 2 million eggs in her ovaries is left at puberty with 300,000 to 500,000. Of these, she will use only 400 to 500 throughout her reproductive life. Until recently, scientists believed that women cease egg production forever at menopause. Intriguing new findings in mice, however, suggest that ovarian stem cells continue to produce eggs throughout the animals' lives. If the same is true in humans, it could have profound implications for infertility and aging in women.

The ovaries housing the eggs are two almond-shaped organs, one on either side of the pelvis in the lower abdomen (see Figure 2.1), attached to the uterus by ligaments. At any one time, **oocytes** may be maturing in the

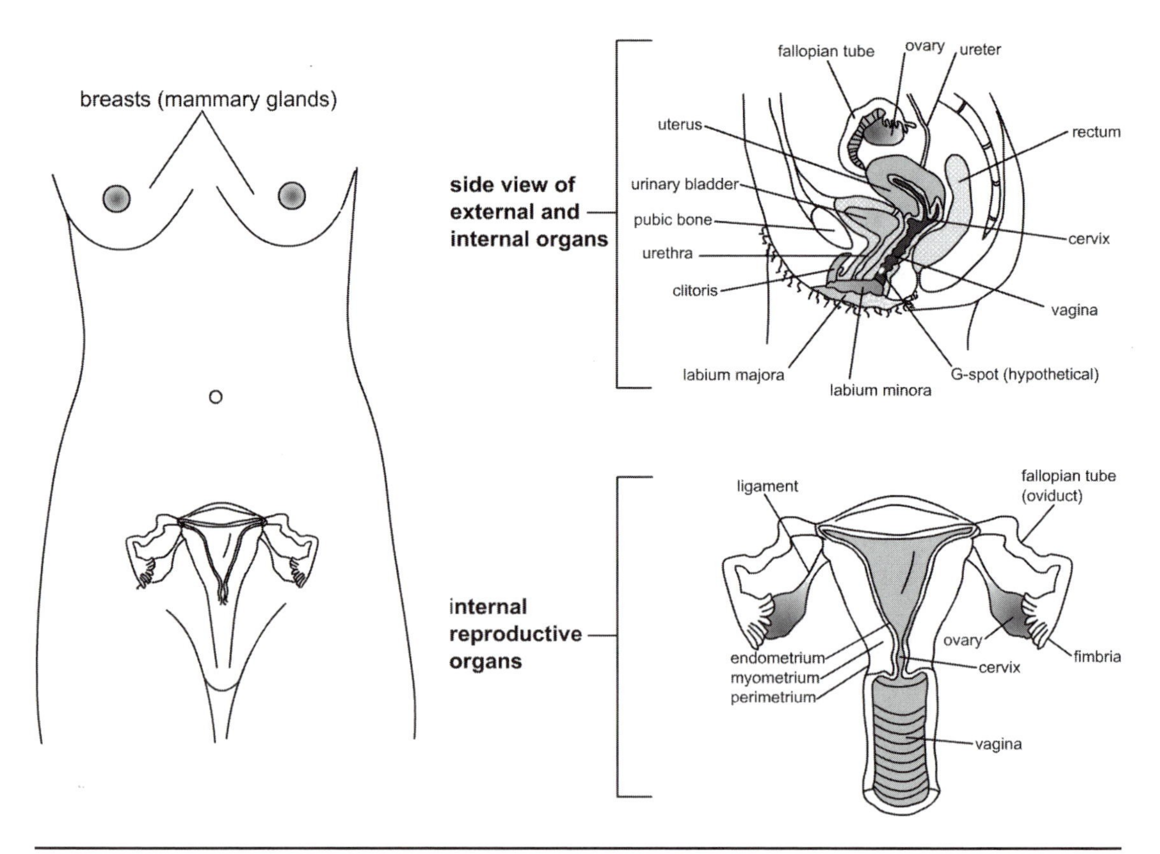

Figure 2.1. Female internal reproductive organs.

ovaries in preparation for eventual fertilization. Regulated by hormones, follicles form around each of the oocytes, one of which matures into a Graafian follicle containing what is by then a secondary oocyte. At ovulation (see Figure 2.2), the mature ovum bursts from the ovary and the next follicle begins to mature. Once the egg is released, it is drawn into the Fallopian tube, or oviduct, that leads into the uterus. Each oviduct, about 5 inches (12.7 centimeters) long, has fringelike projections called *fimbria* that help direct the egg. New studies suggest that the egg's location in the Fallopian tube is a slightly warmer area than surrounding tissue, and that sperm have heat sensors to guide them to the site. If the egg is fertilized, it completes the meiotic division that was arrested at ovulation. Fine hairs called *cilia* coordinate with the contractions of the oviduct to propel the egg down to the uterus, or womb, the upside-down pear-shaped organ that will house the developing fetus. If the egg is not fertilized, the uterus sheds its endometrium, a special lining that thickens in preparation for the fertilized ovum, and expels it during menstruation along with the egg. Powerfully

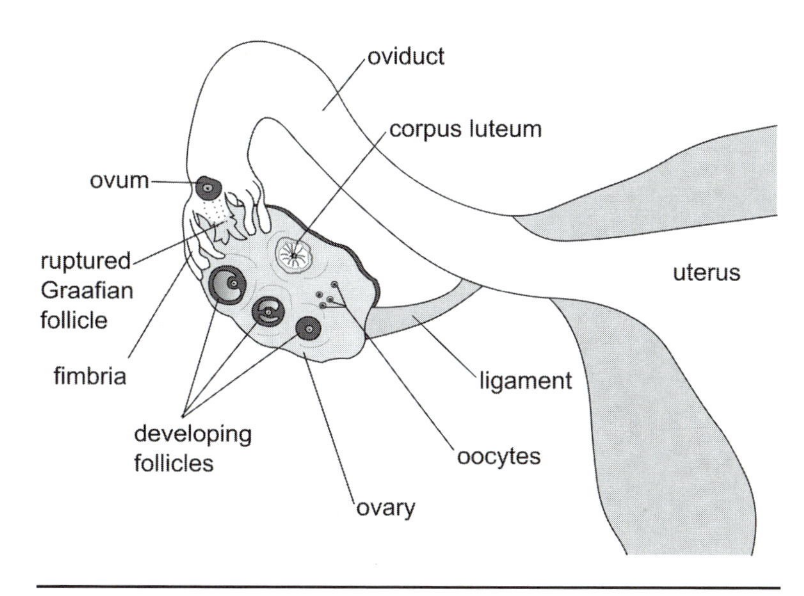

Figure 2.2. Ovulation.

muscular, the uterus is capable of expanding to the size of a basketball during pregnancy. Its wall has three principal layers: the endometrium; the myometrium, the complex of muscles that surrounds the uterus to contract in childbirth and to help reduce uterine size afterwards; and the perimetrium, the outer layer of connective tissue.

The lower third of the uterus narrows into the cervix and the vagina. An organ about one inch in circumference, the cervix protrudes into the upper cavity of the vagina and dilates during childbirth to permit the infant's head to emerge. Its opening into the vagina, the external os, is covered with cervical mucus that changes with the cycling of the menstrual cycle. During pregnancy, it thickens, forming a plug to keep out threatening pathogens. Also known as the birth canal, the vagina is a muscular cavity 3 to 5 inches (7.6 to 12.7 centimeters) long that is ordinarily narrow and collapsed on itself, but it expands and lubricates to admit the erect penis during intercourse. If pregnancy occurs, it stretches much more to accommodate the fetus during childbirth.

While it is not technically part of the genitalia, the urethra (see Figure 2.3) is so intimately associated with both the female and the male reproductive tracts that it merits a brief description here. In the female, urine produced by the kidneys flows down the **ureters** into the bladder, which stores and then releases it through the urethra to exit the body at the **vestibule**. (This will be covered in greater detail in the Urinary System volume of this series.)

There are hidden but important glands on either side of the vagina called

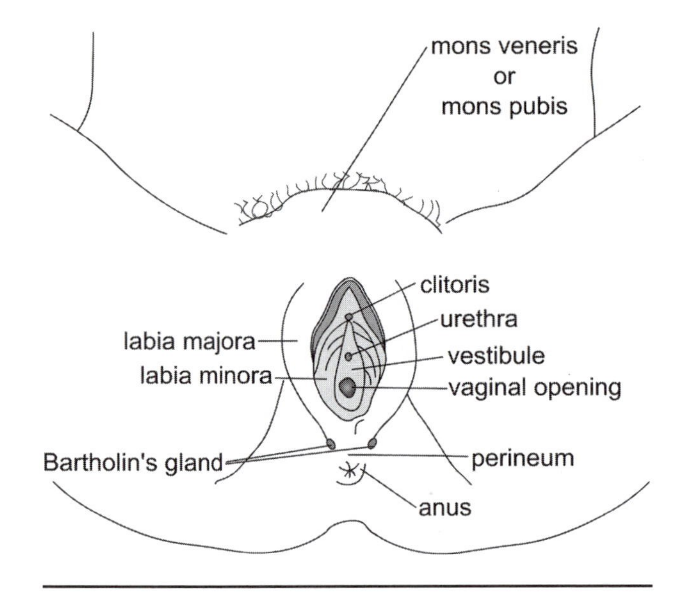

Figure 2.3. Female external reproductive organs.

the *lesser* and *greater vestibular glands* (Bartholin's glands) that secrete fluids into the labia to help lubricate them during sexual arousal. There are two pairs of labia; the labia majora, or larger lips, are the fleshy outer lips that cover the more sensitive, hairless labia minora, or smaller lips. Just behind the urethra, tucked within the labia, is the vaginal opening, which in young girls is partially covered by a thin membrane called the *hymen*. Although this tissue is usually present in those who have never had intercourse, some girls are born without it or unknowingly tear it during physical activity or with the repeated use of tampons. The small expanse of skin between the vaginal opening and anus is called the *perineum*.

Each female is born with an organ devoted entirely to sexual pleasure, the clitoris. Comparable to the penis in terms of sensation, its glans is exquisitely sensitive. Hooded by a prepuce (a membranous, protective tissue capable of sliding back or retracting), the glans is the tip of the clitoris that can be seen at the upper junction of the labia minora. The rest of its body is the shaft, which disappears into the pelvis and consists of two cavities that fill with blood during sexual excitement. There is some evidence of a knot of tissue known as the Gräfenberg spot or "G spot," a controversial area because many believe it does not even exist. Reputed to be another site of female pleasure, it is said to be located an inch or so into the vagina on the front wall, adjacent to the urethra and bladder and behind the pubic bone.

The external female genitals are referred to collectively as either the *vulva* or the *pudendum*. Both terms include the mons veneris or mons pubis, the mound of tissue lying over the pubic bone that in sexually mature women is covered with pubic hair.

The breasts (see Figure 2.4) are mammary glands that, despite their undeniable sexual significance, evolved primarily to feed the young. Even the nipples' so-called erectile tissue—although susceptible to stimulation—is altogether different from that of the penis or clitoris. Smooth muscles within the areola, the differently colored tissue surrounding the nipples, are responsible for the erectility, which probably helps infants find and grasp the nipples more easily.

When she enters puberty, a young girl's breasts begin to accumulate fat and grow. A radiating series of lobes in each breast that reduces to smaller lobules and ends in tiny sacs called *alveoli* drain into lactiferous ducts carrying milk to the nipples. During pregnancy, high levels of estrogen and progesterone cause the alveoli to fill with proteins and other fluids. When the infant is born, the breasts release these nutrients, or **colostrum**, which impart the mother's immunities to the newborn. Within just a few more days, the breasts begin to lactate (produce milk). Nursing not only feeds the child; it also triggers uterine contractions in the mother that help reduce bleeding and recondition uterine muscles stretched by pregnancy.

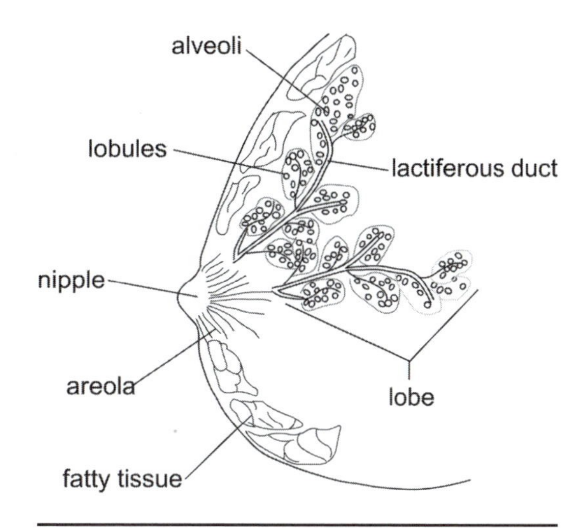

Figure 2.4. Mammary (breast) gland cross section.

When ovulation occurs, the Graafian follicle from which the secondary oocyte erupts is transformed into a corpus luteum, a body that produces the hormones estrogen and progesterone to prepare the uterus for implantation of the fertilized egg (see Figure 2.5). These hormones, under the direction of the pituitary gland, orchestrate the phases of a woman's monthly menstrual cycle. At mid-cycle, ovulation, the lining of the uterus thickens and

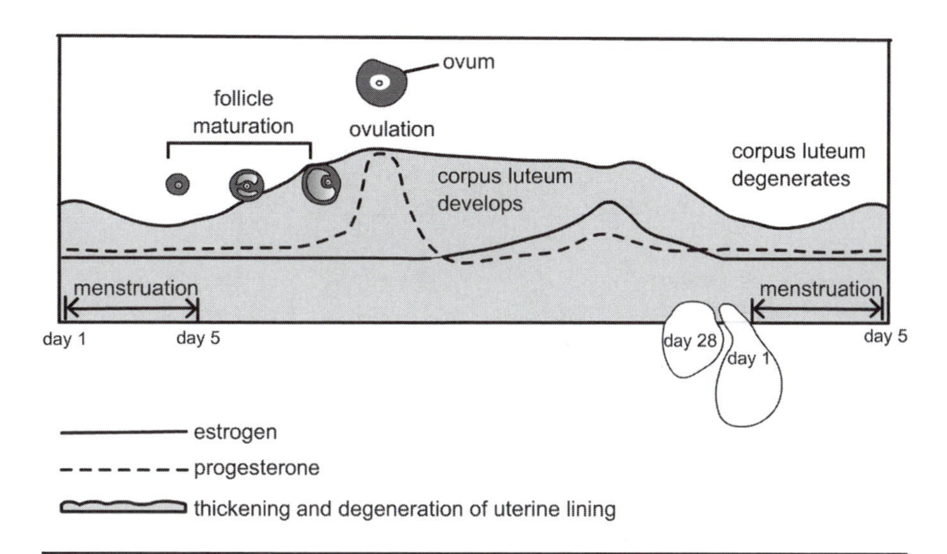

Figure 2.5. Hormone fluctuations during menstrual cycle.

new blood vessels grow. Ten to fourteen days afterwards, if the egg is not fertilized, the corpus luteum ceases hormone production and degenerates; in response, the endometrium breaks down and is expelled from the body in the form of menstrual blood and tissue. The cycle, twenty-eight days long on average, begins anew with the maturation of another primary oocyte in one of the ovaries. Although nature's choice of which ovary releases the ovum each month is entirely random, evidence indicates an equal distribution of labor—each ovary seems to contribute about half of the total eggs ovulated during a woman's reproductive life.

MALE REPRODUCTIVE ORGANS

Spermatogonia, or immature sperm, also undergo changes before they are capable of fertilizing an egg. After they have migrated to the testicles, they reside in some 800 feet (24,000 centimeters) of seminiferous tubules (see Figure 2.6), subsequently dividing by meiosis to become spermatids. Then, under the direction of specialized **Sertoli cells**, they differentiate into spermatozoa, or sperm, several hundred million of which will be made in the testes daily from puberty onward. Since sperm must be kept at the right temperature to remain alive and healthy, the testes are nestled within a pouch of skin, the scrotum, suspended outside the body. In cold weather, the scrotum contracts to pull the testicles closer to the body for warmth. Too much heat can also damage sperm; tight clothing that restricts air cir-

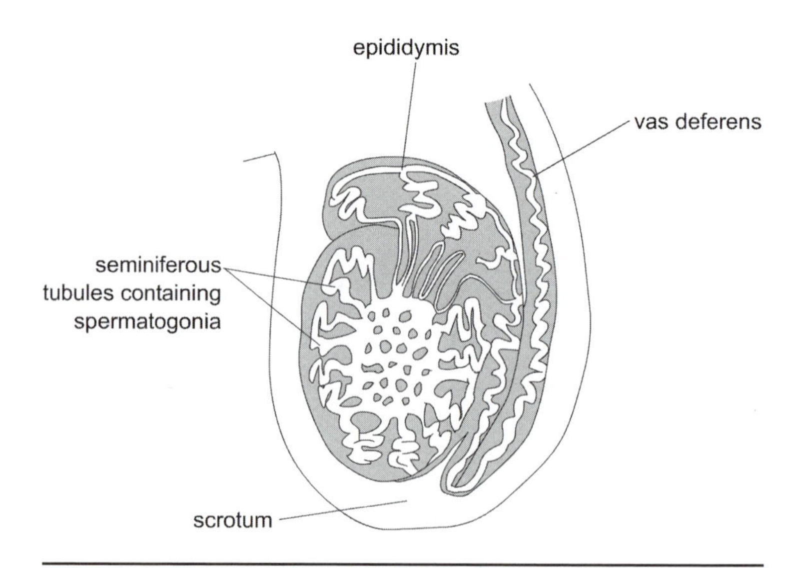

Figure 2.6. Testicle cross section.

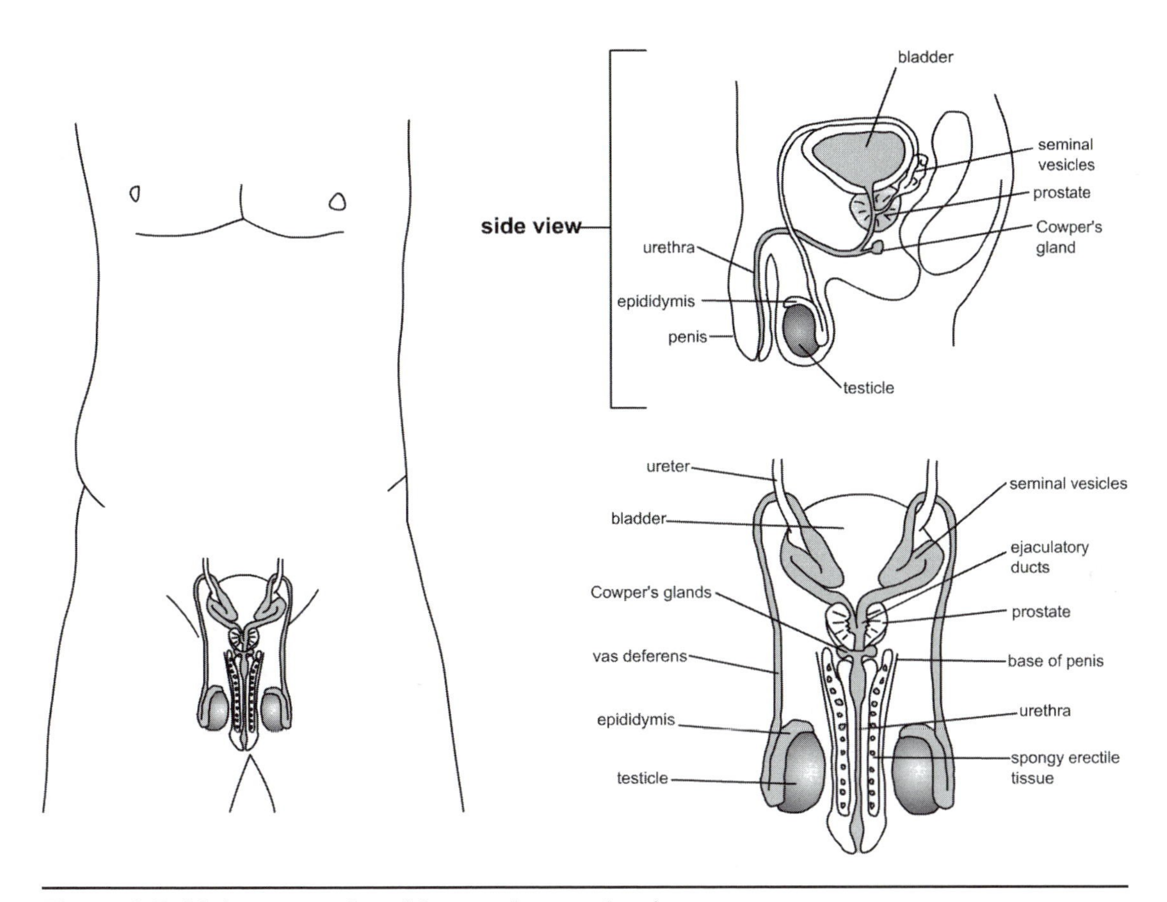

Figure 2.7. Male external and internal reproductive organs.

culation or regular bathing in water that is too hot can kill enough sperm to imperil fertilization.

Manufactured in the testes, sperm mature in storage ducts called the *epididymides.* As ejaculation begins, they enter the vasa deferentia (singular form: vas deferens) that extend up into the body and behind the bladder. There they meet the seminal vesicles, glandular structures that narrow with the vasa deferentia into ureters, and form ejaculatory ducts (see Figure 2.7). Sitting just below the bladder and surrounding the ureters is the prostate, a gland that contracts during ejaculation and secretes fluids into the urethra. This fluid combines with sperm and the secretions of the seminal vesicles and the bulbourethral, or Cowper's, glands to make semen, the whitish fluid ejaculated through the urethra at the end of the penis.

Like the clitoris, the penis contains spongy tissue whose vessels becomes engorged with blood during sexual arousal; this causes an erection that allows the penis to penetrate the vagina and deposit semen during ejaculation. Semen protects sperm in several ways: it provides a safe and fluid environment, it helps neutralize harmful acids in the male urethra and female vagina, and it supplies sperm with needed energy to swim up and into the uterus, a journey that must be completed within twelve to seventy-two hours before they die. Although millions of sperm are contained in a teaspoon or two of semen (the average amount in each ejaculate), fewer than one thousand will reach the Fallopian tubes.

CIRCUMCISION

Males and females alike are born with a prepuce that covers the glans of the penis and clitoris, respectively. In males, the prepuce is also called the *foreskin*. It is attached to the glans but usually, by the time boys are of school age, the foreskin has naturally separated from the penis except for one anchoring point at the frenulum from which it can retract. Many males undergo a procedure to remove the foreskin surgically at birth; this is known as *circumcision*.

Male Circumcision

Male circumcision was first performed centuries before the beginning of the common era (BCE), and many cultures since have continued the practice for hygienic, cultural, or religious reasons. Although the procedure is performed nowadays in a clean environment by a trained person, there is a great deal of controversy surrounding it. Because circumcision usually takes place with no anesthesia when one is an infant and has no say in the matter, and because the absence of a foreskin may deprive men of significant sexual pleasure, many regard it as an unnecessarily cruel procedure akin to mutilation. On the other hand, there is evidence that uncircumcised men are more vulnerable to urinary tract infections and sexually transmitted diseases (STDs) than their circumcised counterparts, and they bear a slightly increased risk of penile cancer as well.

Female Circumcision

Within certain societies around the world, females are circumcised too, usually to ensure their virginity and as part of initiation, religious, or other cultural rites. The procedure is most often performed when girls are three to four years old but, unlike male circumcision, it is not performed under optimum conditions, but rather by midwives or untrained persons using crude implements like razor blades, pieces of glass, or tin can lids. The circumcision is one of three types: the prepuce of the clitoris and sometimes part of the clitoris itself is removed; a portion of the clitoris is removed along with the labia majora and labia minora; or the entire clitoris is removed (clitoridectomy) along with the adjacent labia. Called *infibulation*, this last procedure also involves slicing away the openings of the vagina and urethra and crudely sewing the wounded edges together, sometimes with thorns. A small opening is left for menstrual fluid and urine to escape, but often the woman must be cut open, sometimes brutally, for intercourse.

Cultural pressures supporting female circumcision are so strong and so ingrained in certain areas that the very women it victimized may nevertheless insist that it be performed on their daughters, thus ensuring their children will suffer the same pain, difficult childbirth, and prolonged infections that they themselves suffered. What compounds this tragedy is the spread of the acquired immunodeficiency syndrome (AIDS) virus from contaminated instruments used to perform the circumcision. For these and many other reasons, most cultures reject female genital mutilation as barbaric, and numerous international groups are working to eradicate the practice.

THE NERVOUS SYSTEM

The Brain and Spinal Cord

The brain is the most important sexual organ in the human body. Partnered with the spinal cord to make up the central nervous system, it not only supervises the nervous and the hormonal, or endocrine, systems that regulate the physiology of sexual response (what happens chemically and physically in the body), it is also the repository of the images, thoughts, and feelings that humans associate with sex. (A more detailed description can be found in the Nervous System volume in this series.)

The oldest neural tissue in the human brain, the reptilian brain, is mediated by the influence of mankind's thinking brain, or neocortex. A third area, the limbic brain, may be thought of as a bridge between the two, and the place where emotions and sexual impulses reign. The brain is much more complicated than this; brain structures are coordinated in unaccountably complex ways to balance human instincts and emotions with appropriate behavior. There is nevertheless compelling evidence that some aspects of sexuality are "wired" into the brain. Evolution has etched into humans' **collective unconscious** many of the characteristics that are desirable in a mate, and these traits, unbeknown to the individuals involved, subtly influence how they choose one another. Men may look for the "right" proportions in women, a certain waist-to-hip ratio that suggests a suitable physique for pregnancy and sufficient fat deposits to nourish the young. Women, on the other hand, may tend to seek men who are tall, perhaps because their increased height once conferred hunting advantages and thus meant they would be better providers for the family. In most cases, men and women are not consciously aware of making these choices; they know only that they find someone sexually attractive.

The Peripheral Nervous System and Neurotransmitters

The first arm of the nervous system, the brain and spinal cord combination, works in concert with a second arm, the peripheral nervous system, made up of neural networks that thread throughout the body. It is divided into two parts, the somatic and the autonomic. The somatic sends sensory signals to the brain and activates certain kinds of motor activity but, in human reproduction, the autonomic nervous system (ANS) is the main player. It is always working, orchestrating involuntary functions outside of one's control. Very simply put, its sympathetic system makes preparations for the body to go into action, and its parasympathetic system reverses the preparations. (A third part of the ANS is the enteric system, which influences certain functions like digestion.) When sexual arousal occurs, the sympathetic system increases pulse rate and saliva production, quickens breathing, and raises blood pressure; as excitement ebbs, the parasympathetic system slows pulse rate and breathing, lowers blood pressure, and decreases glandular secretions.

But this is by no means the whole story. Communication among the cells of the body is an extraordinarily complex process that relies on chemical messengers. Some of these are neurotransmitters, made by nerve cells to carry messages in the form of impulses across the synapses, or gaps, that separate them. They convey emotions, thoughts, ideas—all the neural processing that animates humans—and help regulate the secretion of hormones. Some are called the "feel-good" neurotransmitters: dopamine, acetylcholine, and the endorphins, known as natural painkillers. These, along with other biochemicals like norepinephrine and serotonin that *act* as neurotransmitters in the brain, seem to have major roles in sexuality and reproductive functions, probably due to their regulatory effects on mood. An imbalance or deficiency in any one of these neurotransmitters can result in depression, lethargy, insomnia, anxiety, or difficulty with concentration. The body carefully governs this delicate mix primarily through the hypothalamus. The hypothalamus is a regulatory organ in the brain that maintains conditions in the body like temperature, metabolism, and blood pressure, and oversees the limbic system, where emotions like aggression and rage reside. It also issues critical instructions to the pituitary, the master gland that, directly or indirectly, dispenses the body's hormones.

HORMONES

Hormones are another kind of biochemical messenger. There are two types, steroid and nonsteroid, that are secreted by glands throughout the body (see Figure 2.8) that comprise the endocrine and **exocrine systems**. By means of a biofeedback mechanism (see "The Hypothalamus and Mecha-

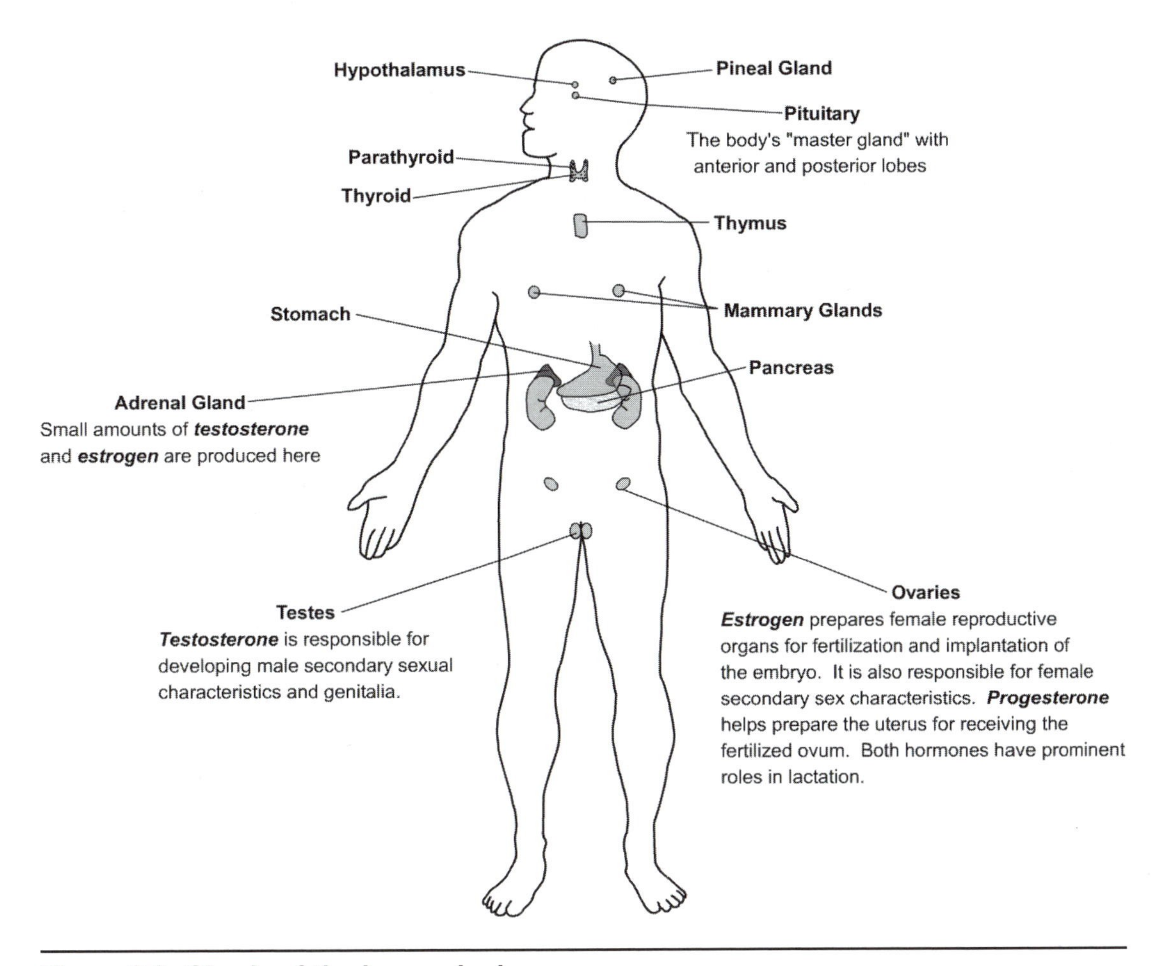

Figure 2.8. Glands of the human body.

nisms of Feedback"), the hypothalamus is alerted to abnormal biochemical levels; in response, it secretes **neurohormones** to tell the anterior and posterior lobes of the pituitary to increase or decrease the relevant hormones. For example, when the hypothalamus releases the neurohormone gonadotropin-releasing hormone (GnRH) to the anterior pituitary, it is telling the gland to produce both follicle-stimulating hormone (FSH) and luteinizing hormone (LH), chemicals that the ovaries and the testes require to support egg and sperm development. Once the pituitary delivers the message, the ovaries and testes begin to do the work they've been hormonally assigned. An example of a hormone produced by the posterior pituitary is oxytocin, related to uterine contractions, milk production, and emotional bonding.

Steroid hormones are made by the adrenal glands in both males and fe-

The Hypothalamus and Mechanisms of Feedback

Also known as the body's "guardian," the hypothalamus ensures **homeostasis**, or equilibrium, within the body. It provides a link between the nervous and **endocrine systems** to regulate temperature; maintain appropriate hormone levels; signal sensations of hunger, fullness, and thirst; juggle aggression, fear, and rage; and set circadian (sleeping versus waking) rhythms. In reproduction, it triggers the pituitary to prompt sex hormone production by the ovaries and testes.

While it is not completely understood how the hypothalamus receives all of its input, it is known that several **afferent** pathways carry neurotransmitters and biochemicals into the organ from the brain and body. To help regulate reproductive functions, it relies on a feedback system; negative feedback is a function of hormone levels. For example, reduced blood concentrations of estrogen and progesterone that result from the degeneration of the corpus luteum after ovulation alert the hypothalamus to a low level of these essential hormones. The hypothalamus releases GnRH that instructs the pituitary to secrete FSH and LH to stimulate ovarian production. As blood levels of the newly produced hormones then rise, the hypothalamus stops triggering their secretion. This mechanism is illustrated in Figure 2.9.

males. They are converted primarily to estrogen in women and to a lesser degree to testosterone; in men, they are converted primarily to testosterone and to a lesser degree to estrogen. These, along with progesterone, are called the *sex hormones* because they are also produced in the ovaries and testes and are fundamental to reproductive biology.

The Role of Hormones in Embryonic Sex Differentiation

When fertilization occurs, the chromosomal arrangement of the merged gametes determines the **genetic sex** of the child. An X ovum fusing with a Y sperm yields an XY embryo, or a male. An X ovum fusing with an X sperm yields a girl. But the "default" embryo—the newly conceived embryo that has not yet developed sexual organs or produced hormones—is generally considered female until other steps involved in sex differentiation take place. It is critical that each step occurs at the right time, or normal development will go awry.

A 7-week-old XY embryo already has testicular tissue identifying him, at the level of his gonads, as male, just as an XX embryo has ovarian tissue identifying her as female. Both embryos have two ducts, the Wolffian and

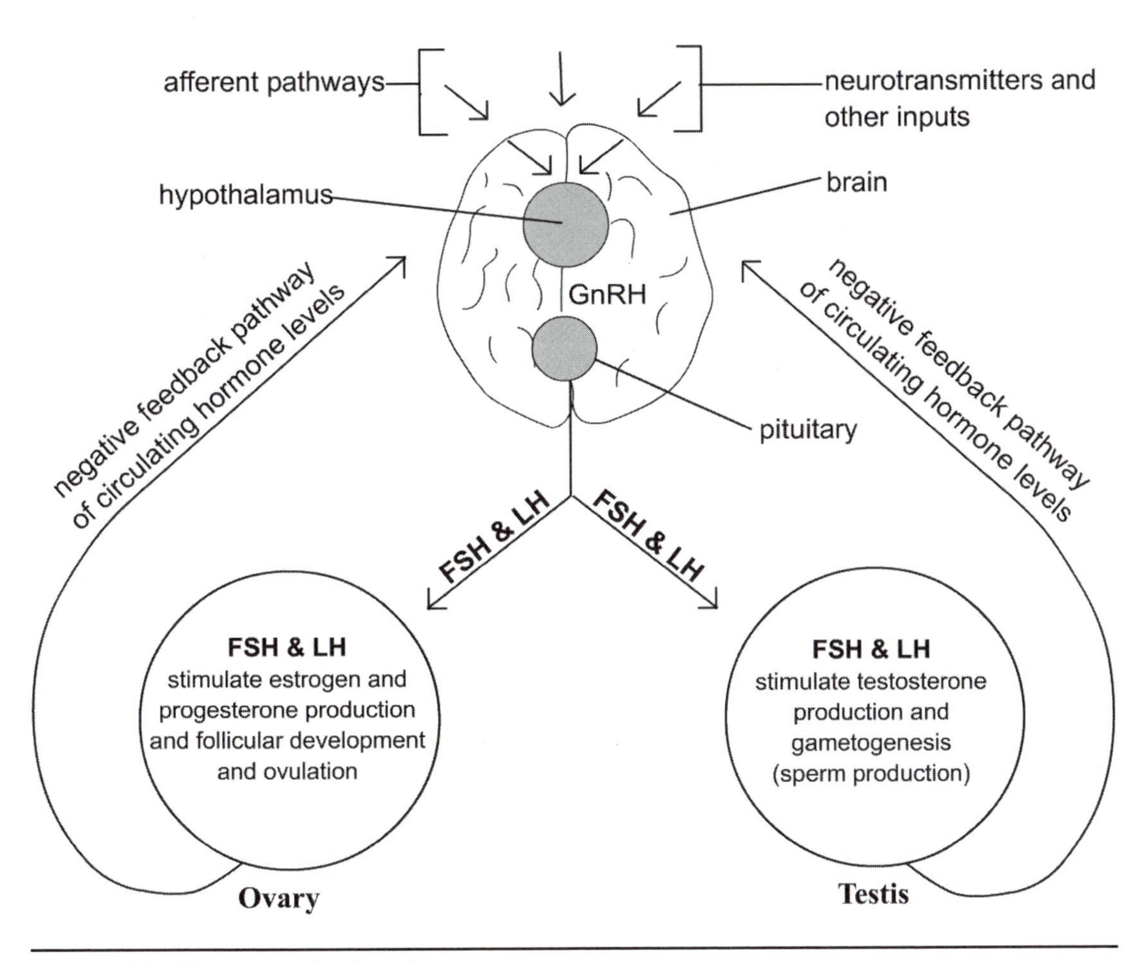

Figure 2.9. Biomechanical pathways.

the Müllerian, that will be transformed into the reproductive structures of either the male or the female, respectively. Scientists believe that a gene expressed only in the embryonic male brain triggers the development of testes; the early testicular tissue secretes testosterone, the transforming agent that causes the Müllerian duct to degenerate and prompts the male genitourinary tract to develop. In contrast, without that testosterone acting on the XX embryo, the Wolffian duct degenerates, leaving the Müllerian to give rise to the female's reproductive organs.

Just as hormones are associated with the differentiation of the gonads and external reproductive organs, so are they associated with sex differentiation in the brain. Once again, testosterone plays an especially critical role.

Remember that the brain of the embryo is still that of the "default" gender, even though male sexual organs may have begun development. To masculinize the brain, a certain amount of estrogen (testosterone that has been converted in the male's body) must pass through the brain barrier. It is not entirely clear why it is the primarily "female" hormone estrogen that is required, but one thing is very clear—estrogen must not be permitted to penetrate the female's brain barrier at this stage. A protein the female embryo produces called alpha-fetoprotein erects a barrier of its own that successfully blocks the estrogen's entry.

This process does not mean that estrogen is responsible for every "male" thought a man possesses. While it seems likely that hormones regulating sex differentiation are responsible for some of the structural differences between the male and female brain, it is also clear that social or cultural factors affect its fundamental development as well. During their youth, boys and girls are bombarded with input that will shape their individual sexual identities. General attitudes as well as individual experiences can influence them; today's "ideal" woman is thin, for instance, but before World War I, with a few exceptions, the plumper, rounded female figure was widely favored. However, there is consensus that even if a given input has no intrinsic sexual content, it could be integrated differently by a male brain than by a female brain, simply because their neurological networks are different. Analyzing what this means in terms of human behavior is not the focus of this book, but the implications offer a tantalizing glimpse into how profoundly the brain affects human sexuality and how differently it shapes male and female perceptions. Chapter 4 discusses how these differences become especially pronounced when children reach puberty.

THE LIBIDO AND THE PHYSIOLOGY OF SEXUAL RESPONSE

As Chapter 1 mentions, the sex drive, or libido, is one of the most compelling drives in human experience, in part because it ensures continuation of the species. It is a source of great pleasure and fulfillment but can also be responsible for great heartbreak and loss—even, in extreme cases, violence. People under its influence have jeopardized marriages, parental rights, job security, economic status, national or international prestige, and countless other symbols of stability and respectability for what is sometimes only a "one-night fling."

Nevertheless, people have sought for centuries to control and heighten libido with aphrodisiacs, including powdered rhinoceros horns, Siberian ginseng, and extracts of yohimbe bark or saw palmetto. In addition to these so-called supplements, there are drugs such as Viagra (legal by prescription), alcohol (also legal, but chemically a depressant and therefore counterproductive), and marijuana (illegal). And then there are the nonconsumables like candlelight and music, money and power, erotic magazines and movies, and a host of others. None of these is truly an aphrodisiac—if, in fact, one exists at all—but there are those who claim some to be effective, at least to a degree. It is not clear if this is due to their inherent properties, the power of suggestion, or both.

Hormones, however, can be effective and are sometimes prescribed, particularly in older people, to supplement those the body no longer produces in quantity. Testosterone is most often touted as *the* sex hormone; menopausal women with flagging libidos whose ovaries have shut down hormone production are increasingly being treated with it. Estrogen too plays an important role, albeit a slightly different one, in that it helps preserve soft skin, lubricates female genitals, maintains arterial elasticity, and protects against the bone loss of **osteoporosis**. Estrogen is not normally associated with increased libido, although there is strong evidence that a woman's sexual desire peaks around ovulation, when her estrogen levels are high.

Despite testosterone's reputation as a libido booster, there have been very recent studies suggesting that the hormone may not deserve so much credit. Evidence shows that nitric oxide, a gas produced naturally in the body, acts like a neurotransmitter and increases blood flow in such a way that it helps induce and sustain male excitement and erection. In women, another biochemical is attracting new attention. Called the vasoactive intestinal peptide (VIP), it increases blood flow to various parts of the body and aids in the lubrication of the vagina. Unlike the more common neurochemicals, VIP is made by a different kind of nerve cell in the body and, like nitric oxide, is believed to act by triggering initial arousal, which is then supported and sustained by the rest of the neural-hormonal network.

Pheromones, biochemicals the sweat glands exude, are thought to enhance men's and women's mutual attraction. There is a great deal of evidence that mammals—including, probably, *Homo sapiens*—and their ancestors have an organ known as the vomeronasal organ (VNO) that senses, or perhaps smells, the pheromone messages emanating from the opposite sex. Also called the "erotic nose," it is a knob of tissue in the nasal cavity that may be responsible for the "instant chemistry" two individuals share. Certainly the evolutionary record of the animal kingdom establishes the validity of this remarkable organ and there is every reason to believe that humans have retained it. But either way, there seems little doubt that people respond sexually, although unconsciously, to the chemical signals that others, just as unconsciously, send their way.

When All Systems Say "Go"

Four phases characterize the complete sexual experience in men and women. These phases are not rigidly distinct from one another; sexual response depends on the harmonious collaboration of different systems and the biochemicals they produce. But they are common reference points used to classify physiological levels of response.

The first phase, excitement, is the period when sexual stimulation activates the sympathetic nervous system to increase heart rate, produce hormonal secretions, and develop tissue swelling and erection. When someone responds to stimuli, the sympathetic arm of the ANS instructs the Cowper's glands in the male to secrete fluid; it tells the arterioles, the blood vessels that carry blood from the heart, to dilate, opening them wider to allow greater blood flow. At the same time, it orders the venules, small veins carrying blood back to the heart, to constrict and thus prevent too much blood from returning; in this way, blood accumulates in the genitals and, in the male, engorges spongy tissue in his penis to cause an erection. Similarly, in the female, engorgement causes her clitoris to become erect while her vestibular glands secrete lubricating fluid and her pulse rate and blood pressure rise. In addition, her breasts and the outer portion of her vagina may swell as her labia become darker in color.

The second phase, plateau, is an intensification of excitement; vascular congestion in the genitals is at its peak, flushing of the skin might occur, and muscles of the thighs and buttocks tighten. Interestingly, the clitoris retracts under its prepuce in this phase and shortens by up to 50 percent, a phenomenon that seems to signal the onset of female orgasm.

The third phase is orgasm, or climax. The male reaches ejaculatory inevitability or "the point of no return" at which he can no longer delay powerful spasms that originate in the epididymides and pulse through the vasa deferentia and prostate. The entire nervous system becomes involved; this is the height of his pleasure and the moment that sexual tension is discharged amid strong contractions that ordinarily lead to ejaculation. But not always. Men may also have a "dry orgasm," an orgasm with no expulsion of seminal fluid. This syndrome can accompany certain neurological conditions or diabetes, it can occur in perfectly healthy males who have ejaculated frequently over a recent period of time, and, most commonly, it happens to young boys nearing puberty when they are experiencing sexual pleasure but their semen-producing organs have not fully matured. Retrograde ejaculation, when semen backs up into the bladder, is another reason for dry orgasm. In the absence of disease, this is a normal phenomenon following several episodes of arousal that do not result in orgasm; fluids quite literally back up into the bladder and subsequently must be released during urination.

Physiologically, the female's orgasm is similar to the male's in terms of buildup and the point at which contractions begin. These are overwhelmingly pleasurable, pulsing through her uterus and vagina in wavelike patterns. Although her orgasm is not necessary for fertilization, uterine spasms in the outer third of the vagina can dip the cervix into contact with sperm and may even help draw sperm up into the cervix, increasing the chance of fertilization. Because she produces no sperm, however, her pleasure and orgasm are almost incidental to the biology of human reproduction; if her menstrual cycle deems her ready for fertilization, there is every possibility she will be impregnated even as a result of rape, no matter how brutal her experience.

Some schools of thought hold that there are two types of female orgasm, vaginal and clitoral, the latter, according to Freudian psychoanalytic theory, being the more "immature," and that an individual woman is capable of experiencing only one kind. Others maintain that the same woman, depending in part on how and where she is stimulated, can experience both. Some women report that an orgasm felt deeply in the vagina is the more intense of the two, triggering stronger uterine and vaginal contractions, while other women say that sensations arising from clitoral stimulation are more pleasurable. All must agree, however, that it is actually the brain that reigns supreme in the sexual response department, for it is only the brain that can trigger a spontaneous orgasm. A sleeping person receiving none of the physical or external stimulation on which arousal normally relies can have an orgasm simply because his or her brain created some exciting imagery. In fact, so complete is the brain's mastery over arousal that some women can fantasize to orgasm while fully conscious, using nothing for stimulation but the rich material her neural activity evokes.

The final period of a sexual encounter is called *resolution*, when the parasympathetic system begins to reverse the excitement that its counterpart ignited and allows blood from engorged genitals to ebb away. Norepinephrine, one of the feel-good hormones, is released, adding to the overall sense of well-being and relaxation that the participants enjoy.

SUMMARY

Because fertilization and fetal development occur inside the body, humans have developed special organs to accommodate these functions. But anatomical capabilities, as marvelous as they are, are only half of the story. Once chromosomes have established genetic differentiation, a delicate balance of biochemicals completes the differentiation of the gonads, the reproductive organs, and the brain. In the sexually mature individual, some of these same chemicals are integral to arousal and libido, fueling the sensations and emotions that drive human sexuality. As the next chapter shows, they are essential to the reproductive events that sex initiates—conception, gestation, and childbirth.

3

From Sex to Pregnancy to Childbirth—Or Not

Despite scientists' sophisticated understanding of human reproductive biology, conception, the beginning of life, cannot be defined. Just because a sperm has penetrated an ovum does not mean fertilization has taken place; that occurs only when the pronuclei of the sperm and egg (two gametes, each with twenty-three chromosomes) fuse to create a zygote, whose nucleus will then contain forty-six chromosomes.

Some view fertilization as the moment of conception. Others feel that fertilization merely launches cell division, producing a nondescript cluster of cells that is not alive until it implants in the womb, establishes a nourishing blood supply, and begins to differentiate into human tissue; these events, they believe, represent conception. So to avoid any misunderstanding, fertilization, rather than conception, is used here to describe the beginning stages of embryogenesis.

Most biological processes build on one another and balance complex cellular and hormonal activity with split-second timing. Nowhere is this more evident than in a human being's first three months in the womb. Developments during the last six months are much less dramatic but are equally important, because they represent organ maturation and growth necessary for life outside the mother's body.

Every parent hopes for a normal pregnancy that ends with the birth of a healthy child, and this chapter describes how such a pregnancy might unfold in the context of a hospital or home delivery attended by physicians or midwives. In addition, prenatal care, fetal and maternal screening, and frequent medical issues pregnant women encounter are discussed.

Many women choose not to have children. In recent decades, thanks to dependable contraceptives and safe, legal abortion, a decision to postpone or avoid parenthood entirely can easily be made. The chapter concludes with a description of the methods used to prevent or terminate unwanted pregnancies.

THE FIRST TRIMESTER

Fertilization

Given the long and dangerous journey each sperm makes, from the upper vault of the vagina through the cervix and uterus and into the Fallopian tubes where it encounters the ovum, it is surprising that fertilization occurs at all. But out of the millions of sperm ejaculated during intercourse, a few hundred or so do indeed survive the journey and, in as little as fifteen minutes or as much as seventy-two hours, reach the upper Fallopian tubes where the ovum awaits, abundantly covered with sperm receptors (see Figure 3.1). Normally only one sperm can penetrate the ovum; the moment its surface enzymes digest a path through the ovum's outer layer, the **zona pellucida**, the ovum dispenses enzymes of its own that break down its receptors and harden its outer layer to make it impenetrable. In a rare event known as **polyspermy**, more than one sperm breaks through. Because this leads to altered chromosome numbers that would result in abnormal development, the ovum fails to develop and is expelled (see "The Question of Twins").

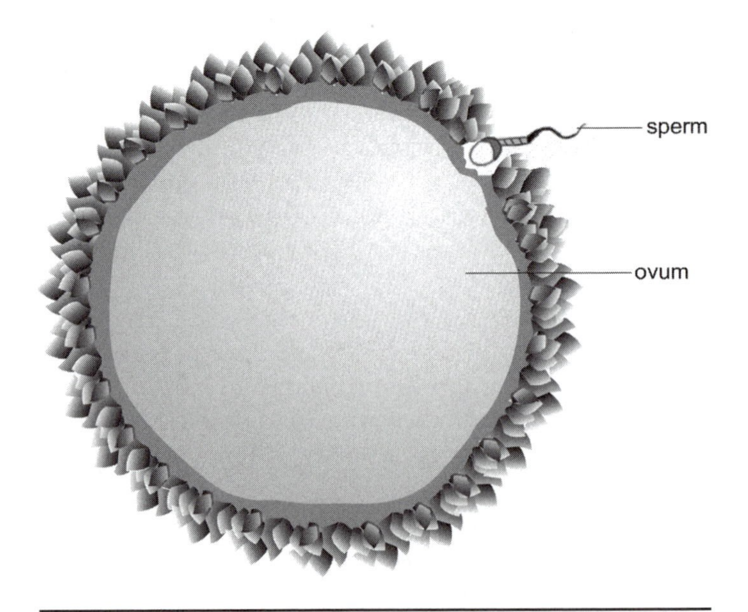

Figure 3.1. Fertilization.

The Question of Twins

Polyspermy raises the question of twins: if only one sperm can fertilize an egg, how do twins develop? The two-fold answer lies in the difference between fraternal and identical twins.

FRATERNAL TWINS

These children are not identical and may be male and female; they result from two different eggs being fertilized by two different sperm. In a given month, a female may ovulate from both ovaries; each of her two Fallopian tubes contains a mature ovum awaiting fertilization. After intercourse, a different sperm fertilizes each. So two genetically distinct siblings—brother-brother, sister-sister, or brother-sister—will become embryos. They are no different from any other sibling combination except that they were nurtured in the womb at the same time rather than, typically, a few years apart. Fraternal twins are often referred to as double-egg twins, and these are the types of twins that tend to run in families.

IDENTICAL TWINS

These two, on the other hand, developed from the same egg and the same sperm and therefore have the same genetic makeup. How do two (or, in the case of triplets or quadruplets, three or four) identical children emerge from one egg? It is because a single fertilized egg, when it begins to divide by mitosis and grow, produces daughter cells identical to itself; rather than staying attached to one another in a single ball, they may split into two (or more) groups. Each group has the same genetic makeup and each begins to grow at the same rate. Two groups of daughter cells produce identical twins, three groups produce identical triplets, and so on. Many mistakenly believe that identical twins can be male and female. Their identical genes clearly make this impossible. Moreover, this so-called "single-egg" twinning does not appear to run in families.

What many may not realize is that twins are not truly identical, even though they have just been described as such. If the twins implant in the uterus at different places, they will receive substances from the mother's bloodstream differently; one twin may be exposed to a greater concentration of certain bacteria, for example, than the other. Their genetic makeup can change slightly after the blastocyst splits into two groups, because a few of one twin's genes may be damaged during subsequent cell division while the other's are not. After birth, they are exposed to environmental influences that will affect the genetic makeup of each twin differently. And then there is genetic imprinting (see Chapter 6), the process by which certain genes are turned "on" or "off" based on which parent they came from; because this somewhat random gene activation will occur differently in each twin, their tissues—some imprinted, some not—will develop somewhat differently.

For this reason, many scientists prefer **monozygotic** to "identical," and nowhere is the reason for this preference better illustrated than in forensic investigation. If a monozygotic twin is suspected of a crime, DNA evidence cannot identify which twin is the culprit because current analytic techniques cannot discern the tiny mutations and variations that differentiate the two. So for now, investigators must hope the guilty twin left some fingerprints behind because, surprisingly, the fingertips of monozygotic twins are not the same. In an excellent example of how environment literally shapes development, the different whorl patterns of identical twins' fingerprints arise in part from the amniotic fluid surrounding them as well as their contact with each other, themselves, and the walls of the uterus.

Although fertilization occurs just after ovulation, the medical community usually calculates pregnancy, or gestation, to begin at the start of a woman's most recent period and to last 280 days (rather than the 266 days elapsing between fertilization and childbirth). The 280 days or forty weeks are divided into three trimesters, the first spanning the weeks between the last menstrual period through the end of the twelfth week. The second trimester is almost four months, covering the thirteenth week through the twenty-seventh; and the third occurs between the twenty-eighth week and the fortieth.

Another system for marking pregnancy's passage, used during the ten-week-long first trimester, pinpoints twenty-three stages of embryogenesis beginning with fertilization at Stage 1, quickly followed by cleavage at Stage 2 (see Figure 3.2 for a timeline of the twenty-three stages).

Cleavage

Fertilized at the upper end of the Fallopian tube, the zygote must make its way to the uterus, implant, and begin developmental growth. As it migrates, it undergoes cleavage, or cell division. The resulting daughter cells (Figure 3.3) divide every few hours, creating a cluster that is the multicelled embryo. The cells then flatten to form tighter junctions, compacting into a rounded **morula** that enters the uterus about the fifth day after fertilization. At Stage 3, the morula develops fluid at the center and becomes a **blastocyst** (see Figure 3.4). As fluid accumulates, the blastocyst for the first time begins to display signs of cell differentiation: there is an inner cell mass that represents the embryo-to-be, and a discernible outermost collection of trophoblast cells that later contributes to the placenta, an organ that supplies nutrients to the fetus and removes waste products such as carbon dioxide, and the umbilical cord, a network of blood vessels that connects the embryo to the placenta. The fluid-filled center is contained by the amnion, a membrane surrounding the embryo that suspends it in amniotic fluid.

Identical twins can result from cleavage if daughter cells split and implant at different sites in the uterus, each developing its own amnion and placenta (see color illustration at the center of this book). These cases account for about 10 percent of identical twins. Another 70 percent or more are formed when the inner cell mass splits, forming two embryos and two amniotic sacs. Or, both may share an amniotic sac; if the cells fail to separate completely, the embryos will be conjoined (or Siamese) twins. Twins sharing an amniotic sac and placenta are at high risk.

Nearly a week after fertilization, the zygote is ready to implant in a uterus already prepped by the pituitary's hormonal directives. At Stage 4, the cluster of cells erupts from the still-intact zona pellucida to land on the blood-rich endometrium of the uterus, where it burrows aggressively and initiates physiological changes in its maternal host that she will soon be unable to ignore.

Stage	Days	Developmental Events
1	—	Fertilization
2	2 to 3	Cleavage: zygote divides, morula develops
3	4 to 5	Blastocyst forms
4	5 to 6	Blastocyst implants
5	7 to 12	Placenta develops, blood supply becomes established
6	13 to 15	Chorionic villi and primitive streak appear
7	15 to 17	Gastrulation and primitive cellular differentiation occur
8	17 to 19	Notocord arises from mesoderm
9	19 to 21	Somites appear, heart begins to beat, forebrain enlarges
10	22 to 23	Primitive spinal cord and blood vessels develop, digestive organs begin to form
11	23 to 26	Nasal pits become visible, mouth depression appears
12	26 to 30	Embryo curls into a "C" shape, slight budding of arms and legs begins
13	28 to 32	Umbilical cord fully forms, lobes of heart appear, blood circulation is established
14	31 to 35	Bones harden, digestive organs mature, mouth appears
15	35 to 38	Body widens, brain circulation is visible
16	37 to 42	Arms bend at elbow, finger rays are distinct
17	42 to 44	Head is larger and the trunk straightens, lungs develop
18	44 to 48	Nose tip is visible, toe rays appear
19	48 to 51	Trunk elongates, genitals show signs of external development, brain enlarges
20	51 to 53	Muscles appear, ovaries and testes become slightly visible
21	53 to 54	External genitals begin sex differentiation
22	54 to 56	Eyelids and external ear more developed, early organ structures becomes evident
23	56 to 60	Head more rounded, limbs are longer, gender is more distinguishable, facial features become more distinct

Figure 3.2. Embryonic development.

Figure 3.3. Early embryonic cell division.

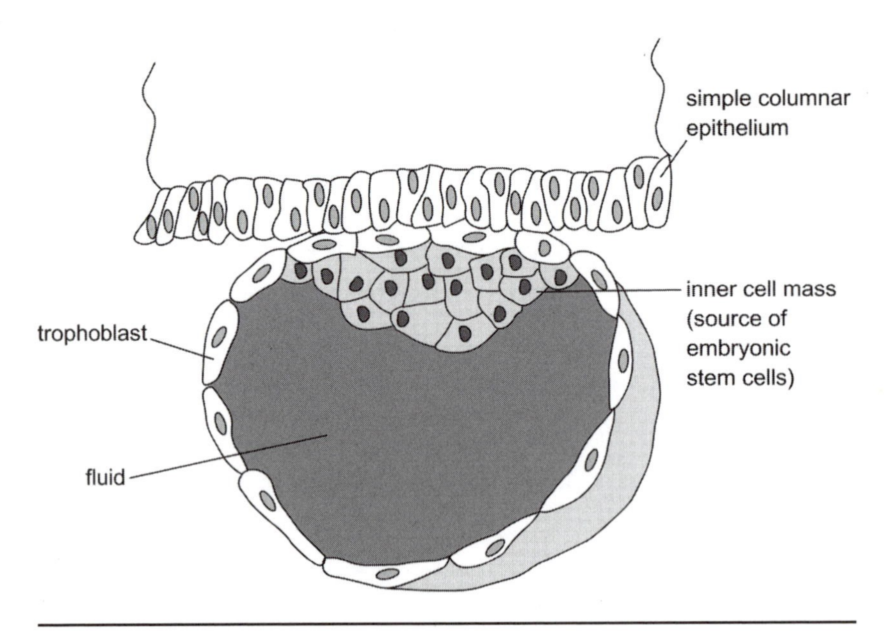

Figure 3.4. Blastocyst.

The Primitive Streak and Gastrulation

About this time, Stages 5 through 8, a placenta forms and becomes anchored to the uterine wall by means of chorionic villi, or "fingers," that infiltrate the endometrium (see Figure 3.5). The blastocyst's inner cell mass divides into an **epiblast** that forms the embryo and a **hypoblast** that forms the yoke sac, a structure that supplies early nutrients in other animal embryos but may be **vestigial** in humans. A groove called the primitive streak forms along the back of the inner cell mass. Cells along this groove migrate inward in a process called *gastrulation*, during which they differentiate into three rudimentary tissues (embryonic germ layers) that in turn give rise to different organs and body systems (see Figure 3.6): the outer ectoderm, cells that will form the nervous system, outer tissues like skin, hair, mouth, and anus, and the lenses of the eyes; the endoderm (sometimes called the gastroderm), an innermost layer that will become the linings of glands and internal organs comprising the digestive, respiratory, and endocrine systems;

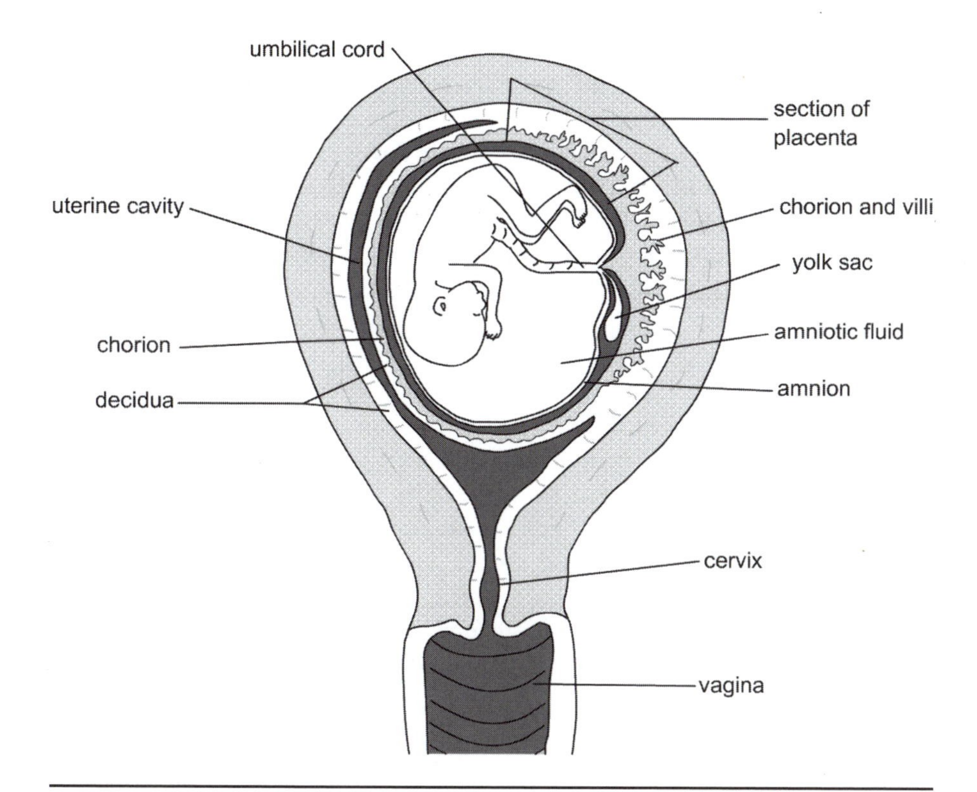

Figure 3.5. Cross section of uterine membranes and tissues during embryogenesis.

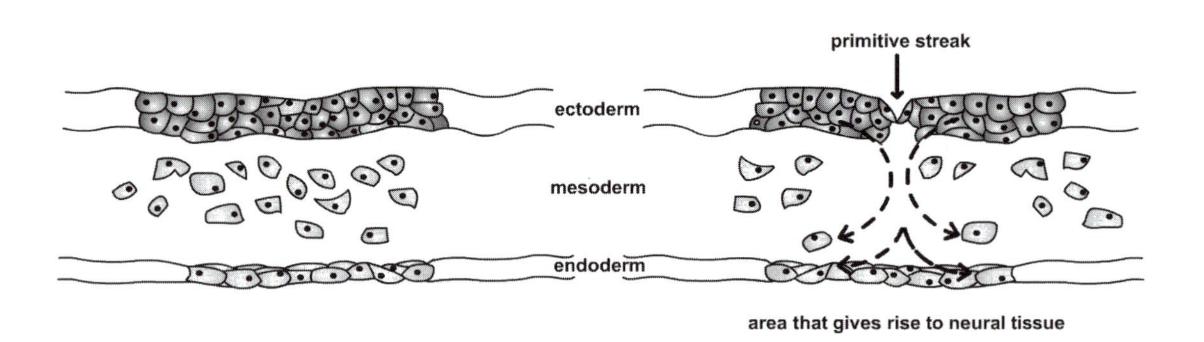

Figure 3.6. Presumptive human gastrulation.
Actual human gastrulation is rarely observed due to ethical constraints on experimenting with human embryos; this representation has been extrapolated from chick and mouse gastrulation.

and the middle layer known as the mesoderm that will develop into the reproductive system, heart, lungs, and blood, and whose cells will develop segmented tissues called somites to become muscles and bones. The notocord, a rodlike structure that orients the organism to top and bottom, front and back, left and right, and forms the basis of the backbone, will also emerge from the mesoderm, followed by primitive development of the nervous system in Stage 8.

Occurring close to the third week of embryogenesis, gastrulation is a pivotal point of development in the first trimester because it marks the beginning of organ formation. This is a critical time for the embryo since certain viruses or drugs or inadequate nutrition can gravely imperil its development and is one of the many reasons prenatal care is so important early in pregnancy.

The Embryo

The embryo's first three months of life are characterized by the most dramatic changes a human organism ever experiences. With a developing notochord serving as an axis, the three cellular layers curl under themselves to form a tube with a hollow "gut" through the middle. The body plan is taking shape and primitive blood vessel networks are developing. Already elongating, the embryo's lumpy body indicates accelerating tissue differentiation. Thickened circles along its upper sides suggest the eyes that are to come and, nearby, puckered areas reveal the sites of future ears. By Stage 9, its primitive heart begins to beat and blood vessels grow, even as a large forebrain starts to dominate the embryonic structure.

In Stages 10 through 12 (weeks 3 and 4), changes become more marked. The embryo curls into a "C" shape, displaying budlike limbs (see Figure 3.7). Digestive organs and glands start to develop, and the early division of the heart into distinct sections begins when blood circulation is established

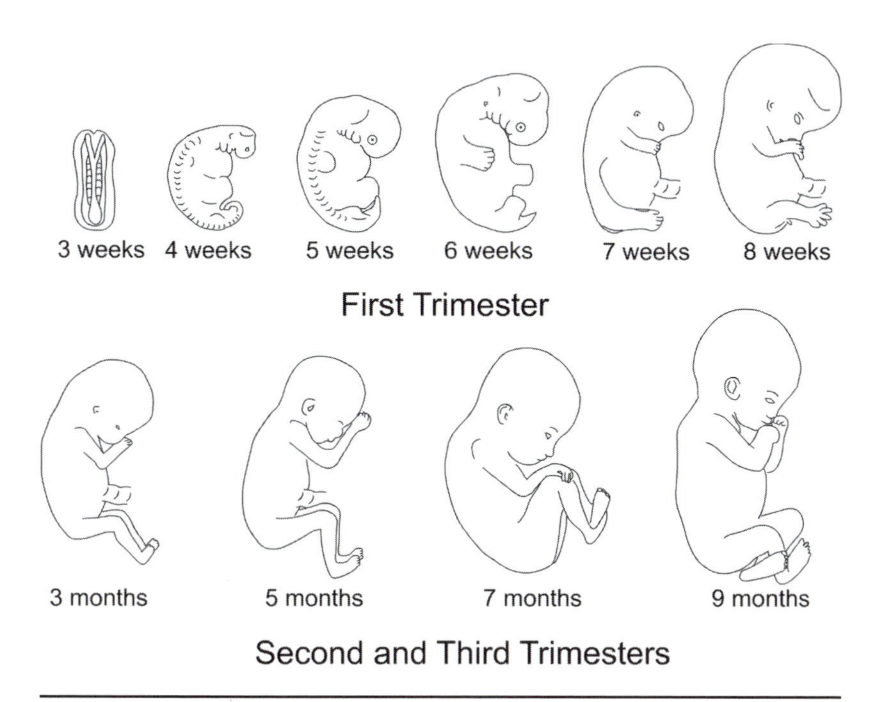

Figure 3.7. Embryonic and fetal development.

and heart valves are more defined. Extensive groundwork for central nervous system development occurs during these stages. A thin layer of skin forms over facial features that are beginning to distinguish themselves; nasal pits are visible and the depression that will become the mouth appears.

Stages 13 through 18, during weeks 5 through 7, are characterized by the appearance of a fully developed umbilical cord, lengthening appendages, and a lobed heart. Digestive organs can be seen now, especially the intestines, which have grown so quickly they temporarily reside outside the abdominal cavity. The esophagus develops out of the trachea and the lungs form. The embryo's trunk straightens to hold its head more erect, and its bones become harder. Fingers and toes are becoming more distinguishable, and genital membranes primed for male or female development reveal the influence of the Wolffian and Müllerian ducts described in Chapter 2.

An increase in brain size, the growth of vocal cords, and progression of organ development characterize the next three weeks of the first trimester, Stages 19 through 23. Male or female genitalia become distinct as the testes begin their slow descent into the scrotum or the ovaries relocate to the pelvic region. The embryo's chin sharpens. The pancreas begins to secrete insulin, hands display dexterity, and both body hair and fingernails grow. By the end of the eighth week, when its facial features are recognizably human, the embryo becomes a fetus. Only one inch (2.54 centimeters) long, it appears to swallow, even hiccup, and in its tiny mouth are thirty-two buds that will become permanent teeth. The muscles function more smoothly and the fetus makes random movements, although its mother cannot yet detect them.

The Mother

It's very unlikely the mother is having any symptoms of her pregnancy this early, but already her body is undergoing radical changes to accommodate a bundle of cells that her body regards as foreign—in this case, the embryo, whose genetic material is different from hers. Human immune systems do not accept alien invasions gracefully; they make very sharp distinctions between "self" and "nonself," and the mother's system would immediately identify this stranger as "nonself."

But just as her system poses a problem, it provides a solution. First, before the fertilized ovum descends all the way down the Fallopian tube, the mother's uterine mucus lining forms a membrane called the *decidua* so that, when the embryo implants, the decidua wraps over the organism to separate it from the inner wall of the uterus. This prevents the mother's immune system from "seeing" the embryo. Second, because there must be some kind of connection between her uterus and her child if the embryo is to survive, a placenta begins to develop from both the decidua and the fetal chorion, a layer of tissue that lies between the embryo and the decidua (Figure 3.5). This becomes the critical passageway for the transfer of nutrients and wastes between mother and child. Life-sustaining chemicals pass back and forth between the two: from the embryo to and from the placenta via the umbilical cord, and from the mother to and from the placenta via its attachment to her uterine wall. This vital activity escapes the surveillance of the immune system because the blood systems of mother and embryo do not touch. How then do they make the exchange? There are complex theories that **genetic imprinting** and **graft rejection** molecules are involved, but to what degree is not known. What is known is that the biochemicals exchanged by mother and fetus rely on the placenta, where **osmosis** and **diffusion** make the transfer.

The placenta also produces hormones, initially just enough to prevent the degeneration of the ovary's corpus luteum that is keeping the endometrial lining richly supplied with blood. When the placenta is capable of producing enough hormones on its own, usually after the first three months, the corpus luteum begins to disintegrate. The yolk sac usually disappears by the seventh or eighth week.

It is remarkable that most of this activity, some of it microscopic, has occurred in just the two weeks since the mother-to-be last ovulated. If she is aware of anything unusual, it may be that her menstrual period is a day or two late, but she could be misled by the light vaginal bleeding or "spotting" that many pregnant women experience during embryonic implantation. (To add to the confusion, some women actually continue to have periods throughout their pregnancy, but this is rare.) Her breasts may be sore too, the kind of tenderness that just precedes her period. So unless she is consciously awaiting confirmation that she is pregnant, she is oblivious to her condition.

That will change by the time her period is four weeks late. The embryo's primitive heart has been beating for two weeks, and the mother will have developed symptoms like fatigue, lightheadedness verging on fainting, and "morning sickness." The latter is a misnomer; the syndrome is more properly known as "pregnancy sickness," or the "nausea and vomiting of pregnancy (NVP)" because the nausea can occur anytime, even in the middle of the night, and may range from slight queasiness to persistent vomiting and dehydration. If the nausea is severe (hyperemesis gravidarum), medical intervention is required, but for the usually mild-to-moderate sickness associated with the first trimester of normal pregnancy, small but frequent meals, avoidance of certain foods (especially those that trigger queasiness; see "Evolution and Pregnancy Sickness"), and plenty of fluids will often alleviate symptoms. Although its causes are not fully understood, the most com-

Evolution and Pregnancy Sickness

Why would nature sanction the nausea and vomiting associated with pregnancy just when women's bodies need to support rapidly growing embryos? This question may be answered by an increasingly popular evolutionary theory of pregnancy sickness that suggests NVP evolved as a protective adaptation to shield the vulnerable embryo from everyday toxins in the human diet. Many of the foods people consume contain parasites or pathogens that are perfectly safe if ingested by fully developed humans, including children. But these same toxins in fragile embryos just beginning to form organs can cause chromosomal damage and other irreversible harm resulting in birth defects or deformities.

Evidence supports this theory. Caffeine, cigarettes, and burnt or barbecued foods that are heavily associated with toxins are notorious for the nausea they induce in newly pregnant women. Certain vegetables in the cabbage family that evolved with a bitter quality to repel pests are poisonous to an embryo and frequently trigger the queasiness of NVP. The most nauseating foods by far are animal-based: eggs, poultry, red meat, and fish, which, unless they are very fresh, are heavily contaminated compared with other foods.

That the nausea and vomiting is worse in the first trimester, when critical organ development occurs, further supports the evolutionary theory. Because it has not been made clear how elevated hormones protect the embryo but it is clear how the mother's avoidance of toxic foods does, it makes sense that the evolutionary theory of NVP is gaining widespread acceptance.

mon hypothesis, at least until recently, pointed to elevated hormones. The reasoning is that elevated hormone levels seem to support a healthier pregnancy; women with healthier pregnancies—those who miscarry less frequently—have more NVP. Therefore, elevated hormones cause NVP. Darwinians—those who look to evolutionary theory to explain biological phenomena—have an entirely different opinion.

Elevated hormones are due in part to the mother's corpus luteum producing more estrogen and progesterone by order of the beta human chorionic gonadotropin (ß-hCG or hCG) hormone. The placenta produces hCG once the embryo embeds in the uterine wall, and the hormone's increased level is one measure by which pregnancy is confirmed. The additional estrogen, which incidentally causes the mother's breasts to become sore, is needed in the development of the embryo's placenta, bones, and sexual characteristics. Progesterone, initially produced by the corpus luteum to build up the endometrial lining before ovulation, normally decreases after ovulation so the lining will disintegrate and be expelled in menstrual fluid. In pregnancy, however, the uterine lining must be sustained, so the progesterone released by the stimulus of the corpus luteum, which in turn is nudged by hCG, ensures that the hormone levels remain high. The woman continues to produce hCG throughout her pregnancy at sufficient levels to prevent uterine contractions, to stimulate breast tissue growth, and to help support the placenta. She also produces human placental lactogen (hPL), both to help her breasts prepare for nursing and to regulate her insulin levels to compensate for some of the nutritional demands the baby is making.

As the mother approaches her second trimester, the fetus, whose basic organs systems are now in place, will begin to make rapid gains in size and weight.

THE SECOND AND THIRD TRIMESTERS

The Fetus

As the child grows taller through the torso (Figure 3.7), its physical features are melding into position; its eyes are moving forward and its ears are close to their permanent location. It can hear, and its reflexes have matured. As the weeks pass and it continues to strengthen, its limbs and torso grow larger to achieve proportion with its head. It is covered in a downy coating called *lanugo* that it will shed before birth. Its heart is pumping more blood and its limbs move rapidly.

Between the sixth and seventh months, the fetus has immature lungs, and only with very sophisticated support could it survive outside the womb. As it develops the capacity to breathe outside the womb, its rapidly growing brain takes over many of its body's functions. Its testes, if it is male, or

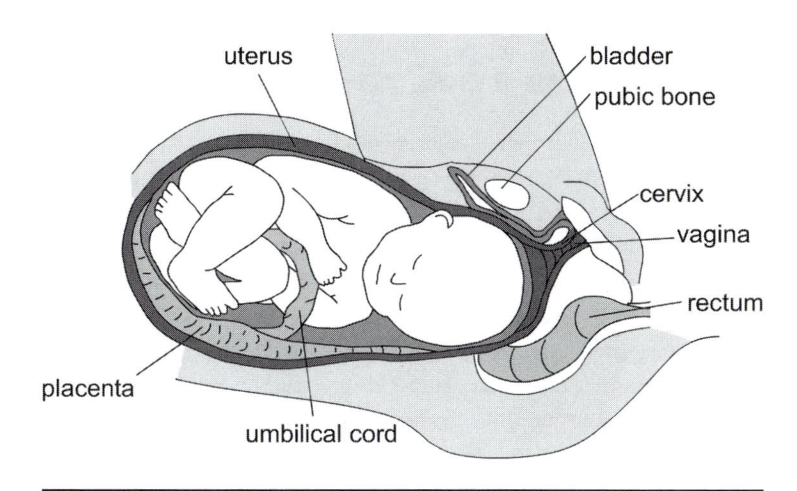

Figure 3.8. Full-term fetus.

ovaries if female, are fully formed, and hair is growing on its head. It is so large now it must bring its knees up into the fetal position to have room in its mother's uterus.

By the ninth month, the fetal lungs have matured and it has built up significant body fat under its skin, which has thickened. Its gastrointestinal system, though immature, is fully functioning, and its nervous system reacts reflexively. Toward the end of the month, when it weighs between 6 and 9 pounds (5.7 and 4.1 kilograms) and is about 20 inches (50.8 centimeters) long, it is ready to be born (see Figure 3.8).

The Mother

A pregnant woman usually feels better during this period than during the first trimester. Nausea and fatigue often vanish early in the fourth month, and she may be suffused with a sense of contentment and well-being. But she may also begin to feel clumsy and overweight.

Growing pressure compresses her bladder, making her urinate frequently; she sometimes has backaches from carrying around the fetus; hormonal changes and the pressure from the baby's growth may have made her constipated and given her hemorrhoids. She might have noticed swollen veins in her legs and stretch marks over her abdomen and breasts.

Colostrum, the nutrient her breasts produce for the child's first **postnatal** day or two, leaks from her breasts, which are sore and tender once more. Unlike the nausea of the first trimester, she might have food cravings that she is compelled to satisfy even though she knows she must not gain too much more weight (see "From Nausea to Cravings"). Headaches are sometimes a problem, as is indigestion (acid reflux) due to the pressure of the

From Nausea to Cravings

In a medical condition called **pica**, some pregnant women compulsively eat laundry starch or other nonfood items, which medical professionals believe reflects an iron or other nutritional deficiency that needs prompt evaluation and treatment. But what about food cravings? Sudden urges for pickles with cheese or olives with dessert are generally harmless and serve as fodder for jokes about the whims of pregnant women. Many simply associate such cravings with hormonal influences and leave it at that.

But perhaps they should not be taken so lightly. Some experts suggest that a strong desire for a particular food is, like the nausea of pregnancy, a message the body sends that shouldn't be ignored. Even though a craving for chocolate seems a long way from a shortage of B vitamins, a relationship between the two has been documented, and some nutritionists suggest that similar relationships between two very different foods often arise in pregnancy. Therefore, they believe, other seemingly benign cravings can be important clues to deficiencies that may need investigation.

baby pushing against the valve that separates her stomach and esophagus. Her uterus is so large now it displaces her diaphragm, making her short of breath, and she has frequent pelvic pain as her bones and joints withstand the growing baby's activity in her womb.

All of these symptoms are normal, although not everyone experiences them. But there are other, more serious, conditions that can arise during pregnancy that require careful medical attention, and this is why **prenatal** care to help avert them is so important (see the later section, "Complications of Pregnancy").

The Father

The father may suffer from a "sympathetic pregnancy." This is a condition in which some fathers-to-be, about 10 percent to 15 percent, undergo physical changes that can be explained only in relation to their partners' pregnancies. Also known as *couvade*, his sympathetic pregnancy may give him nausea, constipation, even food cravings, and his symptoms often coincide, trimester for trimester, with the woman's.

Since hormonal levels are found to fluctuate significantly in those who experience couvade, there is speculation in the scientific community that it is a biological response to some evolutionary imperative to "claim" fatherhood or to evoke deeply paternal instincts so the newborn is doubly protected during its vulnerable infancy. It occurs predominantly in men who live with their pregnant partners, and is thought to be caused by the same pheromones associated with the synchronized menstrual cycles that evolve among women who live together, or the ability of some adoptive mothers to lactate. Although some fathers' symptoms have been reported to extend into the delivery room where they experience severe contractions, their discomfort usually dissipates without incident.

MEDICAL ISSUES SURROUNDING PREGNANCY

Prenatal Care

Any woman contemplating pregnancy should be sure she is in good health, both for her own well-being and for that of the child. She should quit smoking, drinking alcohol, and taking illegal or prescription drugs (unless the latter are approved by her physician) for at least three months before she conceives. If the father is a smoker, he should quit smoking as well to prevent secondhand smoke from endangering the infant.

Once she knows she is pregnant, a woman should see her physician or arrange an appointment at a clinic where healthcare personnel will take her medical history, perform a physical examination, and analyze her urine and blood to determine the state of her overall health. Specific information they seek includes her blood type, Rhesus factor (see the later section, "Complications of Pregnancy"), the presence of infection, anemia, or sexually transmitted diseases, and whether she is immune to the German measles (rubella) virus that endangers normal fetal development. Other risks to the fetus about which pregnant women should be informed include

- the dangers of smoking,
- **fetal alcohol syndrome**,
- parasitic, bacterial, and viral infections, such as **toxoplasmosis**, puerperal fever, or chickenpox, and
- dietary insufficiencies.

The physician or midwife who oversees the pregnancy will set up regular appointments in which to check the mother's weight, blood pressure, and urine, and to feel her abdomen to evaluate the progression of the pregnancy. Under normal conditions, only one or two visits are required in the first trimester. They increase to one visit every four weeks in the second trimester and to one visit every two weeks in the third. From the fourteenth week on, the baby's heartbeat will be monitored as well, and its size checked from about the twentieth week on.

Maternal and Fetal Testing

Several screening and diagnostic procedures are recommended at various stages throughout pregnancy to evaluate the progress of mother and child. Some are routine and others are repeated or introduced when there is reason for concern.

- The **ultrasound scan** is routine and is normally done around twelve weeks to screen for **Down syndrome** and again at eighteen to twenty weeks to verify fetal growth is occurring normally. If the scan captures a view of the external genitalia, it reveals the child's gender.
- A **multiple marker test** done at fifteen to eighteen weeks consists of blood analysis. Additional screening tests are performed in the first and second trimesters if specific complications or disorders are suspected based on the mother's history, symptoms, or the results of the ultrasound; these also detect chromosomal abnormalities and alpha-fetoprotein levels, which can be implicated in Down syndrome or neurological disorders.
- A **glucose tolerance test** will probably be administered to check for gestational diabetes.
- Amniocentesis is a diagnostic procedure usually performed only on mothers age 37 or older because of the small risk of miscarriage it poses. A sample of amniotic fluid is removed from the uterus and examined for fetal cells. These can yield valuable information like the sex of the fetus, its metabolic health, and whether there are chromosomal irregularities indicative of Down syndrome or other disabilities.
- Chorionic villus sampling is a less common diagnostic procedure than amniocentesis but can be done at an earlier stage of fetal development to detect genetic problems or blood abnormalities. This test is normally performed only if specific disorders are suspected.
- Umbilical vein sampling or cordocentesis elicits a great deal of information from fetal blood such as biochemical imbalances that may lead to slowed or retarded development, the presence of infection, and the Rhesus factor. If necessary, the umbilical cord can serve as the route of intrauterine blood transfusion.

Complications of Pregnancy

Despite early detection capabilities and the remarkable sophistication of today's screening and diagnostic tools, not all the complications associated with pregnancy can be avoided, nor can they be successfully treated once they pose a threat.

- Miscarriage and stillbirth are the death of the fetus. Miscarriages usually occur in the first trimester, often because of embryonic abnormalities too severe to allow survival. Stillbirth refers to death of the fetus after twenty-four weeks. Sometimes hormones fail to support the pregnancy, sometimes infections or illnesses compromise it, and sometimes the mother's uterus or cervix have impairments or irregularities that do not allow pregnancy to continue. Although nearly 25 percent of women experience some vaginal bleeding early in pregnancy, only half of those are likely to miscarry.

Bleeding in the second and third trimesters, however, can be very serious and may indicate the placenta is involved.

- Placental separation occurs when the placenta becomes detached from the uterine wall and can have very severe consequences. In the first or second trimester, it means almost certain death of the fetus; in the third trimester, if the extent of the separation is not so great as to place the mother's life in danger from excessive bleeding, a Caesarean section (see "Labor") can usually rescue the fetus.
- Placenta previa is a life-threatening development for the fetus because the placenta lies in the path of the baby through the birth canal and, if it is dislodged during the child's passage, it will cease to provide needed oxygen. Fortunately, the condition can be diagnosed early and, if the mother receives proper bed rest and specialized medical supervision up to her thirty-seventh week, she could deliver a healthy baby by Caesarean section.
- Placental insufficiency is just that; for unknown reasons, the placenta is occasionally unable to transfer nutrients and waste products with enough efficiency to sustain the fetus. Ultrasound imaging and detecting less-than-normal maternal weight gain can lead to a diagnosis of this condition, which may require the induction of labor or a Caesarean section.
- Preeclampsia and eclampsia are very serious conditions with unknown causes. They arise from the placenta and are somehow related to an inadequate blood supply. Preeclampsia, which is symptomless, is detected with blood pressure readings and urine tests during prenatal checkups. It rarely occurs before the twentieth week but, when it does, the only treatment beyond stabilizing the mother is delivery of the fetus and placenta. If preeclampsia develops near term and the mother receives expert medical attention, she may be able to deliver her baby normally. In any event, the condition must be monitored constantly because it could suddenly lead to eclampsia, a worsening of the condition that results in seizures, kidney failure, coma, and, if not treated immediately as the medical emergency it is, death of the baby.
- An ectopic pregnancy means the embryo has implanted in a Fallopian tube where it promptly begins to grow. The discomfort and pain this causes usually alerts the mother by the sixth to tenth week of her pregnancy that something is very wrong. This is the subacute form of ectopic pregnancy in which the Fallopian tube is still intact and the embryo can be destroyed by an injection. The acute form, by contrast, comes on suddenly and is an extreme emergency. The tube ruptures and severe pain, shock, and falling blood pressure imperil the life of the mother. Immediate surgery to remove the tube and the fetus can save her life, but she may have difficulty conceiving again or may even be left infertile.
- Gestational diabetes is diabetes that begins in pregnancy and, in many cases, disappears after the baby is born. Diabetic mothers require careful monitoring because maintaining proper blood sugar levels is critical to fetal health. Often they must deliver by Caesarean section because their children tend to be large; this is due to the increased sugar in the mother's system that crosses the placental barrier and becomes converted in the fetus into larger organs. With careful management, mothers who

develop the disease during pregnancy should expect to deliver healthy children.

- The Rhesus factor (Rh factor) is simply an **antigen** or proteinlike substance that appears in the red blood cells of about 85 percent of people; this makes them Rh positive. The other 15 percent of people are missing the antigen and are Rh negative. It becomes important when the blood of an Rh-negative mother mixes with that of her Rh-positive baby during delivery. No harm is done at that time except that the mother's immune system reacts to the presence of the baby's Rh positive blood by producing anti-Rh-positive antibodies. If she gets pregnant again with an Rh-positive baby, the antibodies her system has produced will seek out and begin to destroy the baby's red blood cells. Monitoring the mother's blood for antibodies during the pregnancy allows physicians to evaluate the fetus and, if it is in distress, transfuse it through the umbilical cord.
- A Group B strep test is usually administered at the thirty-fifth to thirty-seventh week to determine whether the mother is carrying an infection that could be transmitted to her baby. She will also be examined to determine whether the baby is in the normal (head down) or breech (buttocks down) position for delivery, and may have her cervix checked to see if it has dilated in preparation for childbirth.

CHILDBIRTH

Obstetrician vs. Midwife, Hospital vs. Home

Early in her pregnancy, the mother-to-be decides who will provide prenatal care, who will deliver her child, and where the delivery will take place. Often, these decisions are embedded into one—the person providing prenatal care will probably deliver the child at whatever hospital she or he practices—but there are other options, such as home birth, that the mother will want to investigate. Many women, especially those with conditions that put them or their child at risk, choose to see a physician, most likely an obstetrician-gynecologist who specializes in the female reproductive organs and who routinely delivers patients' babies in a hospital. The doctor or other qualified staff handles prenatal care at the physician's office.

In the last few decades, **midwifery** has regained much of the popularity it once enjoyed. Today it is a mainstream alternative to physician-assisted birth, and is frequently chosen because many women appreciate the personalized care midwives extend to their patients. These women feel that many physicians have scheduling priorities that dictate very brief, sometimes impersonal, patient visits, while midwives tend to spend a great deal more time with the mothers whose active participation in the pregnancy they encourage and support. While statistics show that roughly two-thirds of women currently choose obstetricians over midwives, the one-third who opt for midwives report they are extremely happy with their choice. Many women who give birth in hospitals choose to have both an obstetrician and a midwife (or a professional labor companion known as a *doula*) attend them.

Hospital births supervised by obstetricians tend to involve what some regard as unnecessary medical interventions designed to serve the physician more than the mother or the child: **episiotomies**, vacuum extraction of the fetus or use of forceps, artificial induction of labor, and even Caesarean section are the primary examples. Physicians' frequent use of anesthesia (compared to midwife-supervised births), most frequently **epidurals**, can lead to **iatrogenic** complications requiring yet more medical intervention.

On the other hand, the birthing centers that many hospitals feature, offering a cozy atmosphere where the family gathers during labor and where a less "institutional" delivery can take place, soften the sterile atmosphere that characterizes in-hospital childbirth and appeal to many who feel they combine the best of both worlds. If the pregnancy is high risk, or even a normal one presenting last-minute complications, the mother-to-be may feel safer in a hospital where she is in the right hands at the right time.

Home births also offer numerous advantages. Many claim there is no substitute for the privacy, intimacy, and comfort provided by a home birth, and statistics show that in normal pregnancies, their safety rate is equal to that of hospital births. Certified nurse midwives (CNMs), the most highly trained of their profession, are legal in every state, can prescribe certain medications, and have ongoing back-up arrangements with physicians should serious medical issues arise. Midwives' attitude toward childbirth—that it is a natural, healthy, and participatory event the whole family can experience, one that need not be accelerated with artificial inducements—is very attractive to many women and their partners, and no doubt accounts for the rising popularity of midwifery. Although some insurers are slower than others to recognize them, midwives can be affiliated with patients' health maintenance organizations in many areas, and their services are eligible for reimbursement from Medicaid in some states.

For simplicity, "attendants," whether physicians or midwives, is the term used throughout this section to describe the medical personnel standing by at the birth.

Natural Childbirth vs. Childbirth with Anesthesia

New mothers, especially, derive great benefit from childbirth classes. Knowledge of what to expect reduces their anxiety and fear, and, once childbirth begins, equips them to make appropriate choices when they are their most vulnerable. The most well-known methods are taught in Lamaze or Bradley classes, both of which emphasize proper breathing and relaxation to manage pain. Local hospitals sometimes offer their own version of similar classes. In most cases, the mother's partner is encouraged to attend as he or she may act as the coach during labor and delivery, especially if the mother chooses to have natural childbirth—childbirth with no anesthesia.

If she does expect to have anesthesia, the mother-to-be should explore her options. The most common pain reliever is a regional anesthetic known as an epidural or intrathecal delivered via catheters inserted near her spine; these are positioned to numb a region of her body, either the entire lower half or just the abdominal area. It eliminates much of the pain of labor while allowing the mother to be fully aware and in some cases able to walk around. Epidurals can also be used during Caesarean deliveries if the surgery suddenly becomes necessary. Sometimes local anesthesia is used as well to numb a specific part of the body, such as the vaginal area if an episiotomy must be performed.

Another way to manage the pain of labor is with tranquilizers, sedatives, or narcotics that are injected into the mother's vein to induce an overall sense of relaxation. Since these drugs can suppress the respiration and mental activity of both mother and fetus, many physicians as well as the mothers-to-be reject this option in favor of epidurals. General anesthesia, which puts the mother to sleep, is generally reserved for emergency surgical procedures.

Alternatively, women can self-administer a mixture of oxygen and gas through an inhalation mask that helps numb pain centers in the brain. "Natural" pain relief measures include breathing and relaxation techniques, massage, acupuncture, **transcutaneous electrical nerve stimulation (TENS)**, shifting positions, being submerged in water to support the mother's weight, and hypnosis.

Labor

No one knows for sure what triggers labor, although some suggest that the fetus sets things in motion because of its eagerness to escape the confines of the uterus. In the last decade, however, researchers working with sheep—a mammal whose hormonal pathways during pregnancy are similar enough to a human's to provide a good study model—suggest one likely scenario, and, once again, hormones are at the controls. Corticotropin-releasing hormone (CRH), released by the hypothalamus, triggers a cascade of activity that alters the ratio of estrogen and progesterone the placenta produces. The increased estrogen that results nudges the fetal pituitary to produce oxytocin and the mother's uterus to secrete prostaglandin, and it is this oxytocin-prostaglandin combination that sets off contractions. A classic example of a positive feedback hormonal mechanism (see "The Hypothalamus and Mechanisms of Feedback" in Chapter 2), those same contractions set off the secretion of more oxytocin-prostaglandin, which sets off more contractions. So, in a sense, the baby does control labor, or at least the timing of it. Just as the embryonic placenta initiates much of the biochemical management of early pregnancy by secreting hCG, it engineers pregnancy's end by secreting labor-inducing hormones.

The mother-to-be experiences a number of symptoms as her labor begins.

She has a low backache and doesn't "feel right." Her "water" may break—amniotic fluid escapes when the membrane surrounding the fetus ruptures. As her cervix softens to dilate for the passage of the child's head, she may notice a bit of bleeding or expel the mucus plug that once sealed her cervical opening into the uterus. Her early contractions may be of the Braxton-Hicks variety, which are merely tightenings of the uterus that occur in the last weeks of pregnancy.

The onset of real and regular contractions, though fairly mild and ten to fifteen minutes apart, announce that the active phase of Stage 1 labor is about to begin. If she has decided to have the baby at the hospital or birthing center, this is the time to get ready to go there. Otherwise, she should call her midwife and prepare for her home birth.

As her labor progresses, her cervix undergoes **effacement** and **dilation** as stronger and more painful contractions increase in frequency. In nulliparous women—those who have had no children—labor can last as long as fifteen hours, sometimes longer. In multiparous women—women who have had more than one child—it usually lasts about eight hours. As labor progresses and the contractions increase in frequency, the cervix continues to dilate and the mother will feel the urge to push. When the contractions occur at two- to four-minute intervals and the cervix is fully dilated to 4 inches (10 centimeters), the baby is ready to emerge.

Delivery

The fetus moves down the pelvis a little, exerting enough pressure that the mother's urge to push becomes very intense. With each contraction, she pushes the baby down a little further, but with each relaxation it slips back up slightly. If she can resist pushing too much as the baby progresses through the vaginal opening, she will avoid tearing; if not, she may need an episiotomy so the child's head can emerge. Doulas and midwives take pride in their custom of gently lubricating and stretching the vaginal and genital tissues to prevent tearing and episiotomies, but there is often not enough time or personnel to attend to these details in a hospital setting.

As the child's head "crowns"—its widest part appears in the vaginal opening—the mother bears down again as its shoulders and body are carefully guided out (see Figure 3.9). The baby is covered with a layer of cheesy vernix, which kept him waterproof in the womb. The attendants towel him off to stimulate his breathing and keep him warm. As he sucks in oxygen and begins to wail, he is placed on the mother's stomach until her uterus begins to contract again. In her third and final stage of labor, the placenta separates from her uterine wall and she must push once again to expel it (Figure 3.10).

The attendants examine the placenta carefully to be sure it was delivered

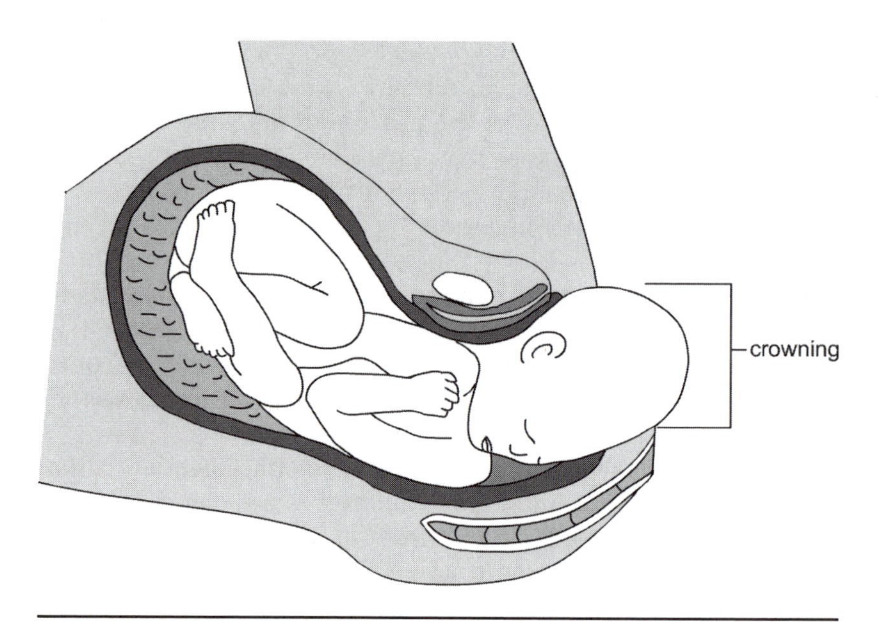

Figure 3.9. Crowning.

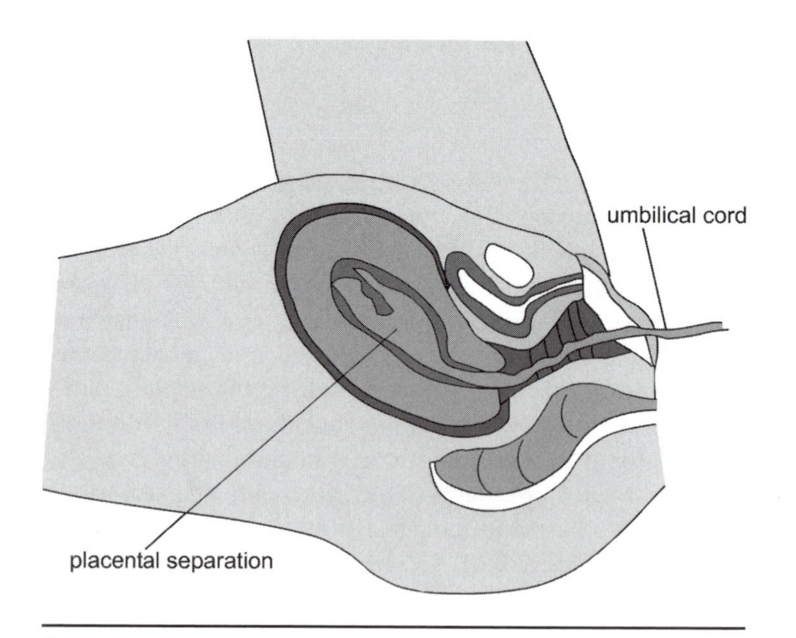

Figure 3.10. Placental separation.

in its entirety, noting that the umbilical cord connecting the infant and placenta is still pulsing with blood and oxygen. They leave it attached for several minutes so the newborn receives maximum benefit from the oxygen and nutrients it contains.

When pregnancy and birth are happily anticipated events, this can be a magical time for new mothers, many of whom feel euphoric despite their fatigue. Others may be overcome by tears of emotion or simply by exhaustion. However, many modern health insurance policies dictate that new mothers leave the hospital forty-eight hours after a normal vaginal birth (ninety-six hours after a Caesarean section), allowing little time for rest.

Complications of Childbirth

Several things can go wrong during childbirth, many of which can be anticipated and prevented by prenatal screening and examination. For anticipated problems, many Western physicians tend to rely on inducing labor or performing Caesarean sections to control the timing and progression of childbirth. Other complications occur spontaneously. Common concerns are:

- Birth injury to fetus. The child's passage through the birth canal can sometimes lead to bruising, nerve damage in the neck, or, if forceps are used, to a temporarily misshapen head.
- Breech or abnormal presentation. If the baby is not head down in the birth canal but presents buttocks first in a breech position, or if the fetus's head is too large for the mother's pelvis, a Caesarean section may be necessary.
- Fetal distress. If the umbilical cord becomes compressed or trapped between the baby and the birth canal, or if circulatory problems arise from other causes, the fetus may be deprived of oxygen.
- Premature delivery. This can be due to several causes and can sometimes be averted with drugs; if not, and if the fetus cannot survive outside the womb on its own, it must be placed in intensive care where sophisticated **neonatal** measures may support its life for the necessary days or weeks of its continuing development.
- Inducing labor. Drugs are used to stimulate labor contractions and speed labor along. It may be necessary because the fetus is in distress or because the mother has a medical condition that necessitates delivery of the child. Labor is also induced when the amniotic membranes have ruptured but labor has not started naturally.
- Caesarean sections. These are most frequently performed when there is fetal distress, when the baby's size or presentation makes vaginal delivery impossible, or when labor and delivery threaten the mother's safety and health. After anesthetizing the mother, usually with an epidural, the physician makes an incision in her abdomen and removes the baby by hand, with the use of forceps, or with vacuum extraction. Recovery from a Caesarean is usually uneventful but it classifies as abdominal surgery. There are two points of view about whether women who have had Caesarean sec-

tions are able to have normal vaginal deliveries in subsequent childbirth. At one time, scar tissue from the first Caesarean was thought to preclude successful vaginal delivery, but many obstetricians today believe vaginal birth after Caesarean (VBAC) to be a viable option, as long the incision from the first Caesarean is not a deterrent in some way.

- Birth, or **congenital**, defects (some of these are discussed in the next chapter).

POSTPARTUM

Breastfeeding

In 1997, the American Academy of Pediatrics issued a policy statement that put forth its position that breastfeeding is beneficial for babies from a developmental standpoint and because it offers some protection against chronic diseases. Other sources confirm that the baby receives nutrition as well as his mother's immunities to help protect it from certain allergies, bacterial infections, and viruses. Suckling at the breast has been shown to enhance an infant's jaw development and lead to improved speech. Rather than limiting a mother's mobility, breastfeeding seems to be a convenience, reducing the need for bulky equipment, refrigeration, and fuel sources for heating formula. Equally important, there is evidence that women who breastfeed are at reduced risk of developing breast and ovarian cancer.

There are disadvantages to the practice as well. Breastfeeding puts constraints on the mother's time, even though breast pumps allow her to store her milk in advance. It can cause a great deal of breast discomfort, it creates a nuisance when breasts "leak" at inappropriate times, and, in certain cultures, it is socially unacceptable to nurse one's child in public. Nursing sustains high levels of prolactin that tends to suppress libido, which many women find distressing. And some women simply do not want to breastfeed or, if they do, may have jobs outside the home that make it very difficult and severely limit the amount of time they can devote to it. This is one reason that, in Western countries, women generally breastfeed for a year or even less, although some may continue longer with nighttime nursing. Other cultures, however, continue to nurse until the children wean themselves, which may not occur until the child is 3 or 4 years old or more. The duration has a great deal to do with prevailing customs, but there is no doubt it also has a great deal to do with economy, convenience, and the pleasurable bonding it can engender between mother and child.

Transition

When a woman's menstrual periods return after childbirth, she may find her first period or two heavy and painful; she is also likely to find her libido has evaporated. In part, this is due to her body's need to recover from childbirth as well as the strain and stress a new baby causes, but those are probably secondary. The primary cause is hormonal: her estrogen and progesterone levels are very low after childbirth and are likely to remain that way for some time, especially if she nurses. Even when breastfeeding is behind her, it may take a long time for her sex hormones to stabilize.

PREVENTING AND TERMINATING PREGNANCY

Since humans first made the connection between sex and pregnancy, they have sought ways to have the pleasure of the former without the consequence of the latter. They have often succeeded, although there were millions of unplanned pregnancies along the way. That does not have to be the case now because there are several birth control methods that, if used properly, are nearly 100 percent effective in preventing pregnancy. Disease is another matter, however (see "Contraceptives and Sexually Transmitted Diseases [STDs]").

Given the differing viewpoints about when life begins, it is important to distinguish between contraceptives and abortifacients. Contraceptives are agents that prevent ovulation, kill sperm, or block it from reaching the ovum. Abortifacients are agents that intervene after fertilization to prevent implantation of the blastocyst in the uterus and cause the embryo to be aborted. For this reason, contraceptives and contraceptives that combine abortifacients are discussed separately in the following sections.

If an unwanted pregnancy does occur, millions of women turn to abortion. The U.S. Supreme Court has ruled that abortion is legal up to the twenty-fourth week of pregnancy, but many states have imposed limits on the procedure. Some require parental notification or consent for minors, oth-

Contraceptives and Sexually Transmitted Diseases (STDs)

With the exception of the condom, no birth control measure or device offers protection against sexually transmitted diseases (STDs), and even the condom cannot guarantee 100 percent protection unless it is used properly each time and every time. In fact, the National Institutes of Health takes the position that condoms can protect against the HIV/AIDS virus and some cases of male gonorrhea, but it is not certain they protect against other STDs.

Microbicides, which are antiviral and antibacterial agents that prevent pregnancy and some STDs, are increasingly in demand, and they are under investigation in research laboratories around the country.

ers require waiting periods, and still others have pushed the date of fetal viability to earlier than twenty-four weeks. New laws restricting abortion and court challenges to that legislation are pending in many states. Because abortion is defined as the premature delivery of a human embryo or fetus that cannot survive outside the womb, the use of abortifacients is sometimes referred to as abortion.

Contraceptives

The famous—or infamous—withdrawal method, or coitus interruptus, is hardly a birth control method at all because it fails so frequently. It amounts to the male withdrawing from the vagina immediately before he ejaculates. But even if he does, the lubricating fluids his glands produce during sex can be loaded with sperm well before he ejaculates.

The rhythm method is a natural birth control measure endorsed by the Catholic Church. (Other natural contraceptive practices like the body temperature method, described in the following paragraph, are presumably allowed as well, but the Church specifically bans any artificial means of contraception.) The rhythm method relies on the partners' avoidance of intercourse at exactly the time of month the ovum would be poised for fertilization in the Fallopian tube. Its effectiveness, if very carefully timed, approaches 80 percent. Like Catholicism, many religions, such as Orthodox Judaism and very traditional factions within Hinduism, Islam, and Christianity, impose restrictions on birth control except in certain cases, while some modern branches of these religions favor a more liberal approach to contraceptive use. In many cultures, opponents of contraception view it as a license for immorality, while proponents view it as essential to worldwide health and avoidance of many threats posed by overpopulation.

The Billings method and body temperature method are techniques in which a woman relies on her body to tell her when she is most fertile. In the Billings method, she carefully observes her vaginal discharge for a few critical days, watching for sticky, opaque cervical mucus that tells her she is ovulating. Some women may need informal training to be sure they can recognize the characteristic discharge. In the body temperature method, she simply takes her temperature during those same critical days to discern the slight elevation that occurs right after ovulation. These methods at best are 80 percent effective.

The cervical cap and diaphragm are "barrier" methods in that they block the entry of sperm into the uterus. The cap fits snugly over the cervix itself; the diaphragm is larger and is placed in the upper vagina in front of the cervix. Both devices must be fitted by a doctor and both should be used with an over-the-counter spermicide to improve their effectiveness. They must be kept in place for several hours after intercourse, then removed and cleaned for reuse before having intercourse again. They are inexpensive and safe to use, although some spermicides irritate genital tissue, and the flexible ring surrounding the diaphragm can put enough pressure on the bladder to cause a uterine infection. Even with proper use, these are only about 80 percent effective, although that rate rises a few points if spermicides are conscientiously used as well.

A new contraceptive technique now on the market is a tiny springlike device implanted into the Fallopian tube near its entrance into the uterus. This blocks sperm from reaching the ovum where fertilization can occur. Considered permanent contraception, it cannot be reversed, so it is primarily aimed at older women who have had their families and want no more children. One advantage is that, unlike tubal ligation (see the later section, "Sterilization"), the device is inserted vaginally through the uterus in a physician's office under local anesthesia. It requires no surgery and recovery from the procedure is almost immediate.

Condoms are thin, flexible sheaths that fit over the erect penis and catch the sperm during ejaculation, blocking it from reaching the cervix. Condoms are effective only if they don't leak and should be checked carefully for cracks or holes before use. They are available in pharmacies and grocery stores everywhere and are often manufactured with a spermicide lubricant or packaged with a printed recommendation to use a spermicide for greater effectiveness.

So far, there are no other male contraceptives on the market except the condom (see the discussion of vasectomy in the later section, "Sterilization"), but research trials are underway to develop a hormonally based agent that suppresses sperm production.

There is also a female condom, a pouch that fits inside the vagina and performs the same function as the male condom. Male condoms are slightly more effective than female, 85 percent versus about 80 percent, respectively.

The spermicidal sponge, foam, cream, jelly, and suppositories are different products with the same mode of action. Any one of them can be inserted into the vagina before intercourse. They are safe and comfortable to use unless either partner is allergic to an ingredient in the spermicide. The sponge must be removed and disposed of after intercourse. Effectiveness can range from 80 percent to 85 percent, but if a woman has been pregnant before, the effectiveness of the sponge is reduced to about 60 percent.

Contraceptive-Abortifacient Combinations

The contraceptive pill, or "the pill" as it's generally known, created a revolution in birth control in the 1960s. It gave women complete control over their reproductive destiny for the first time and, many believe, spawned the sexual revolution of the '60s by freeing women to explore their sexuality exempt from pregnancy. In the forty years since, the pill's estrogen and progestin levels have been decreased to deliver effective contraception with minimal hormones. It works by suppressing ovulation, but has a back-up mechanism, in case the first fails, that irritates the uterine lining to prevent implantation of the embryo. Taken cyclically, it is fairly safe, but risks increase greatly for older women and those who smoke. It is about 95 percent effective.

While the pill must be taken daily, implants are small rods or capsules placed underneath the skin of the upper arm that are fully effective twenty-four hours after insertion. Like the pill, they deliver estrogen and progesterone to suppress oocyte development, prevent implantation, and thicken cervical mucus to interfere with the penetration of sperm. The implants, 95 percent to 99 percent effective in preventing pregnancy, can remain in place for three to five years, at which time they must be replaced.

In 2003, a new birth control pill was introduced that reduces a woman's menstrual period from once a month to once every three months. Although its mode of action is similar to that of the original pill, the active ingredient in the new product is taken for eighty-four days rather than twenty.

Injections given periodically at a doctor's office are convenient, but can cause irritation at the site of injection. Depending on their formulation, they give one to three months' protection that is 99 percent effective.

Abortifacients

The "mini-pill" is another kind of birth control pill, but has only one hormone, progesterone, which inflames the uterine lining and makes it inhospitable to the fertilized ovum. It is about 88 percent to 99 percent effective.

The intrauterine device (IUD) or intrauterine system (IUS) is a small device that a doctor inserts into the uterus. Although some IUDs are treated with a spermicide, their primary mode of action is to irritate the uterine lining. They last for up to five years, are very affordable, and are effective up to 99 percent, but they can cause cramping and heavy bleeding.

The vaginal ring, inserted by the user into the vagina for three weeks, releases hormones that confer 95 percent to 99 percent effectiveness. The ring can be awkward to remove.

The patch, which may cause skin irritation where it is applied on the body, is a hormonal delivery system changed weekly, with three weeks on and one off. It is 95 percent to 99 percent effective.

There is significant research underway to develop a contraceptive vaccine, renewable with "boosters," that harnesses the immune system. Studies are particularly focused on producing antibodies that block the hormones supporting pregnancy or that make sperm and eggs resistant to fusing at fertilization. Much more research is needed to establish the efficacy of these agents and to confirm that normal fertility returns after the antibodies are diminished in the body.

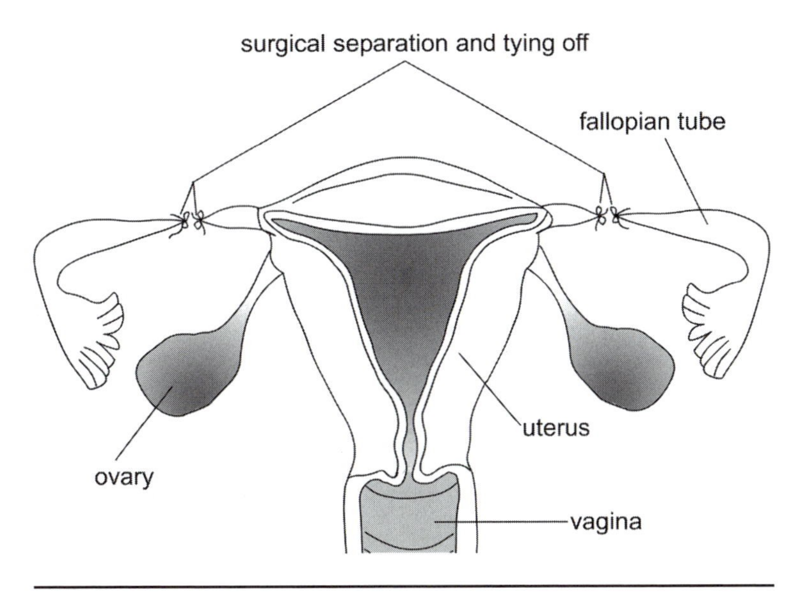

Figure 3.11. Tubal ligation.

Sterilization

Perhaps the most drastic birth control measures are surgical interventions that in most cases leave the patient sterile. In women, tubal ligation (see Figure 3.11) is often performed after she has had all her children, and involves cutting and tying off the Fallopian tubes. In men, vasectomies (see Figure 3.12) are similar; the vasa deferentia are cut and tied off to prevent sperm from leaving the testes. Fortunately, neither of these procedures adversely affects the sex life of either partner. While both surgeries, particularly vasectomy, can occasionally be reversed under special circumstances, patients should expect to be permanently sterilized before they agree to the procedures.

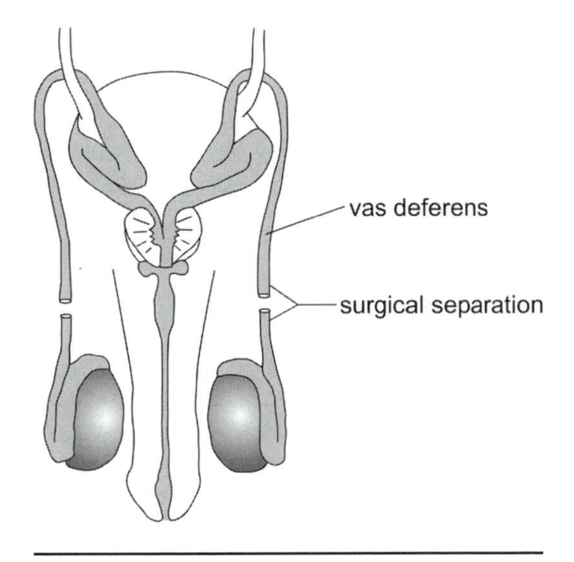

Figure 3.12. Vasectomy.

Abortion

MEDICAL ABORTIONS

The "morning-after" pill, effective only until the forty-ninth day after fertilization, is actually a treatment protocol involving two drugs. RU-486, or mifepristone, the drug contained in the first three pills the patient must take, is a steroid that starts the abortion process by blocking the progesterone providing nourishment to the embryo. The patient takes the three pills at once, followed a few days later by a fourth pill containing another drug, misoprostol, which helps her expel the embryo. The pills can cause nausea, heart attack, serious bleeding, and impair the patient's future fertility. They must not be taken casually, and their use requires several doctor's visits and his or her close supervision.

Methotrexate is another drug that causes a medical abortion and works by destroying embryonic cells. Like RU-486, it must be followed with a dose, or sometimes two doses, of misoprostol to cause the contractions that will expel the embryo.

SURGICAL ABORTIONS

In the first trimester, the most frequent procedure is vacuum abortion performed in a doctor's office or clinic in a single visit. The doctor widens the cervix by inserting a series of tapered rods, then suctions out the contents of the uterus. This procedure should not be performed before the sixth week of pregnancy, to prevent damaging the uterus.

In the second trimester, pregnancy may be terminated by inducing labor. Known as the induction or instillation method, an injection is given in a hospital setting that will cause the pregnant woman to go into labor and expel the fetus a few hours later. Another procedure, usually done with anesthesia on an outpatient basis, is a dilation and evacuation (D&E). This is similar to the aspiration method except that the cervical dilation must be greater and the contents of the uterus, which are larger, must not only be suctioned but scraped out with a **curette** as well. A similar procedure is dilation and curettage (D&C), which may or may not involve suctioning.

Abortion is rarely performed in the third trimester, and then only when the mother has a severe medical problem that prevents her from carrying the child to term, or when there is a gross fetal abnormality that would not permit the child to live. When an abortion must be performed at this stage, it is one of two kinds: dilation and extraction (D&X) or hysterotomy. The former is a partial-birth abortion, so called because the dead fetus is delivered vaginally. Hysterotomy, the alternative procedure, is very similar to a Caesarean section.

Although there are a few mild side effects associated with the contraceptives listed here, the contraceptive-abortifacients and the abortifacients, because they are formulated with hormones, can have side effects ranging from mild to serious and can pose significant risk to certain women. The IUDs and IUSs, because they involve devices that are placed inside the uterus, can cause infection and, if they perforate other organs, serious bleeding and damage. Anyone considering these birth control measures should educate herself about these side effects and risks and discuss her concerns with her healthcare professionals.

SUMMARY

The more people learn about reproductive physiology, the greater their surprise that pregnancy and childbirth succeed as often as they do. The complex demands of pregnancy require extraordinary adaptations on the part of the mother, who must not only accommodate an alien presence for nine months but must also withstand childbirth.

What causes wonder, again and again, is the transformation of a one-celled zygote into a full-term infant. The exquisite precision of embryogenesis awes every parent anew, and the bond that subsequently forms between a mother and her newborn child is, despite its prevalence, no less profound.

For many reasons, some women are persuaded that pregnancy must wait, or they feel it must be terminated. Several contraceptive and abortifacient measures exist today that allow them to exercise their options and manage their reproductive destiny as never before.

4

Sexual Development

The sex differentiation that humans undergo in utero, or within the uterus, takes place on four developmental levels: genetic, gonadal, genital, and cerebral. In most cases, these stages proceed normally and a child is born with the mental and physical conformation that defines him or her as male or female. But sometimes things go wrong that alter a child's sexual physiology or psychology; some of the causes are clear but others, particularly those that lead to variations in the way a person experiences and expresses his or her sexuality, are murky and complex.

This chapter examines both normal childhood sexual development from infancy through adolescence as well as that which is different from the norm. "Normal" is misleading if it is used to imply that any behavior or sexual expression deviating from it is abnormal. The word does not carry that connotation here; instead, it refers to the range of sexual behaviors most frequently observed in a given culture.

By the end of adolescence, sexual identity has taken shape and, for the most part, guides how each individual expresses his or her sexuality during the reproductive years. In middle age, however, when women begin to enter menopause, they are, as in puberty, overwhelmed at times by hormone fluctuations that disrupt their physiological and psychological equilibrium. Although men retain their fertility throughout life, they too experience significant midlife hormonal upheavals. This "change of life," profound though it may be for the individual going through it, is merely biology's acknowledgement that one's reproductive prime is nearing an end.

CHILDREN AND SEXUALITY

Developmental experts understand that a child's sexual behavior has nothing to do with reproduction and everything to do with pleasure and comfort. At first, an infant's rubbing against other objects or absently playing with his genitals may be accidental self-discovery, much as he plays with his nose or ears. In the womb, male fetuses have been reported to have erections and female fetuses to have vaginal lubrication, and there is ultrasound evidence that they randomly stimulate their genitals.

After a child has passed his second birthday, newly developed muscular control allows more coordinated genital stimulation. By the age of three, it is quite common for both girls and boys to masturbate deliberately, and some behavioral scientists report that children have been observed stimulating themselves to what appears, in nearly all respects, to be an orgasm. Many suggest that children this age have already developed a heightened awareness of genital pleasure and satisfaction that helps form the foundation of their adult sexual identities.

A universal and normal activity, masturbation may indicate the child is having some kind of psychological difficulty only if he engages in it to the exclusion of other healthy activities. Parents who disapprove of and punish children for normal masturbation may be creating a greater problem than the one they fear is in the making. To a child, curious self-exploration and the pursuit of an activity solely because it feels good are the most natural behaviors in the world, and child psychology holds that these pursuits have a legitimate, perhaps even crucial, place in development. While parents need to teach children about an appropriate time and place for masturbation, conveying a sense of shame or disgust about the activity itself can be quite damaging.

When children are between 3 and 5 years old, curiosity about their bodies and their absorption in its exploration tend to shift to others. They often engage in playing "doctor" or "I'll show you mine if you show me yours" games to learn more about the anatomy of the opposite sex. In Freudian psychoanalytic theory, this period of development may foster **castration anxiety** in some boys, who view a girl's lack of a penis with alarm and fear, and **penis envy** in some girls, who may associate the male organ with power.

By school age, most children have a well-socialized intuition of appropriate sexual behavior or expression. They start to understand the joke or anatomical references embedded in some adult humor and, by the time they are eight, may explore the bodies of their same-sex friends to satisfy mutual curiosity. Children in this age group tend to disdain contact with the opposite sex, an attitude underlying "no girls" clubhouses that young boys build or the "girls only" informal social events that deny boys entry.

What is central to healthy development is positive emotional and physical closeness between the parent(s) or surrogates and children while they are still very young. As children grow, they begin to transfer good feelings experienced with parents to their friends in the form of hugging and handholding. Those who are deprived of nurturing emotional relationships and healthy physical contact from a very young age may grow up unable to trust, to express affection, or to share intimacy, and are thus destined for a stunted or—depending on the extent of neglect or abuse—even a distorted psychosexual personality (see "Terms Related to Sexuality"). Biology marches on, however, thrusting

Terms Related to Sexuality

Medical, psychological, and cultural issues surrounding sexual orientation and ambiguous genitalia have generated a vocabulary of terms to help sort out much of the confusion they create. A few of these terms are:

- *Gender:* The genetic, gonadal, genital, and cerebral qualities associated with being a male or a female.
- *Gender assignment:* The sex to which one is assigned at birth based on physical appearance; may be the opposite of gender identity, resulting in psychosexual conflict.
- *Gender identity:* Whatever gender an individual feels himself or herself to be.
- *Gonadal aplasia:* Absence of a gonadal organ or gonadal tissue.
- *Gonadal dysgenesis:* Malformation or abnormal development of a gonadal organ or tissue.
- *Intersex:* Individuals who are born with male and female sexual organs.
- *Psychosexual:* Relates to the emotional, psychological, and physical aspects of sex.
- *Sex-change operation:* Surgery performed on those who wish to change anatomically from one gender to another; it is supplemented with hormone treatments.
- *Sexual identity:* How a person perceives his sexual orientation—heterosexual, homosexual, or bisexual; different from gender identity, which refers to whether individuals see themselves as male or female.
- *Sexual orientation:* One person's sexual, social, and emotional attraction to another based on the other individual's gender.
- *Transgenders:* A group of people who, on some level, feel they are a different gender from their anatomical one; this is a very broad category that includes transsexuals, homosexuals, and transvestites (both heterosexual and homosexual).

girls between the ages of 9 and 14 and boys between 10 and 15 into the three- to five-year period known as puberty and transforming them into sexually mature teenagers.

Puberty and Adolescence

Puberty is not quite the same as adolescence since the latter usually refers to all the physical and emotional development of 12- to 20-year-olds, not just the sexual. Although puberty is the focus of this section, it is discussed within the overall framework of adolescence; to explore a child's inauguration into reproductive sexuality isolated from the emotional and intellectual developments that accompany it is a one-dimensional approach that fails to reflect the enormity of the changes he or she undergoes.

For decades it has been thought that the rising levels of sex hormones produced by the testicles and ovaries initiate a series of bodily changes collectively called *puberty*. More recent evidence shows that when children are as young as six or seven, the adrenal glands produce slowly rising levels of dihydroepiandrosterone (DHEA) that initiate biochemical puberty, or adrenarche. It "primes the pump" of sexual development and reportedly accounts for the increased interest in sexual matters that many children display two or more years before they exhibit any physical signs of puberty.

There has been rapidly increasing evidence in recent years that girls are beginning to mature much earlier, in some cases when they are as young as five or six. Numerous studies implicate obesity as the major cause, suggesting that underexercised, overweight girls carry excess body fat that acts like another gland, converting adrenal hormones into estrogen. What is interesting, however, are studies with males that show it is the leaner and taller boys, not the heavier ones, who tend to begin puberty at younger ages. Although the data on females is consistent and verifiable, the data on males continues to puzzle experts.

In pediatric circles, the progression of puberty is marked by the Tanner stages (see "Tanner Stages of Puberty"); each stage is distinguished not by the age at which it occurs but by the level of development. In contrast, Figure 4.1 shows the ages of children at which, on average, **secondary sexual characteristics** appear. Precocious and delayed puberty and other anomalies are discussed at the end of this section.

Physical Development

The brain's hypothalamus and the hormones it recruits direct the arrival of secondary sexual characteristics, just as they do many other physiological changes. Activated by numerous impulses, this supreme regulatory organ releases GnRH, the gonadotropin-releasing hormone that triggers the pituitary to secrete FSH and LH. These are the same chemicals described

Tanner Stages of Puberty

GIRLS, AGES 8 TO 13

Stage 1
Prepubescent

Stage 2
Breast budding; sparse growth of pubic hair; height accelerates

Stage 3
Breasts reach small adult size; pubic hair gets darker and curls, starts to spread; acne may develop

Stage 4
Areola forms secondary mound above the rest of the breast; pubic hair coarsens but is not fully grown in

Stage 5
Adult breasts are developed and pubic hair is fully grown in; no appreciable height increase occurs

BOYS, AGES 9 TO 14

Stage 1
Prepubescent

Stage 2
Scrotum reddens and testes enlarge; sparse growth of pubic hair

Stage 3
Penis grows longer and scrotum enlarges; pubic hair gets darker and curls, starts to spread; height accelerates; voice breaks; breasts may develop (gynecomastia)

Stage 4
Penis continues to enlarge and scrotum darkens; pubic hair coarse but is not fully grown in; acne may develop

Stage 5
Adult pubic hair is fully grown in; facial hair grows in

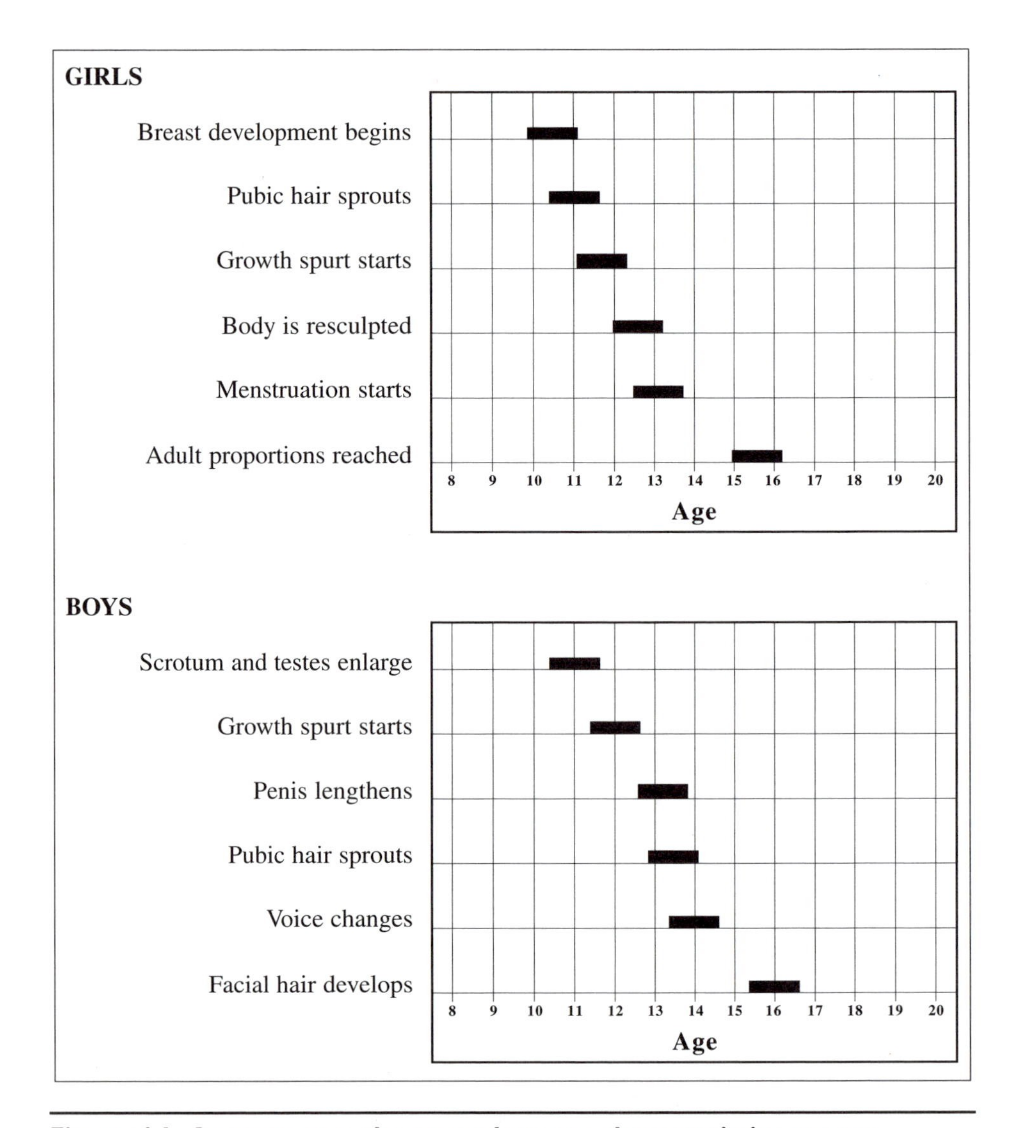

Figure 4.1. Average ages the secondary sex characteristics appear.

in Chapter 2 that stimulate both the male testes and the female ovaries to produce sex hormones, and that is exactly what they do at puberty.

That the changes these hormones produce often occur in different sequences from those described here does not, in most cases, mean there are developmental problems, but simply reflects the fact that the pace of individual growth differs. The events characterizing normal puberty that are common to both boys and girls include changing sleep-wake cycles, increased physical growth, development of sexual organs, and significant psychological and intellectual maturation. Accelerated body growth during the entire period nearly doubles teenagers' weight and increases their height by roughly 25 percent.

FEMALES

A girl of 9 or 10 years old is normally too young to menstruate because her hormone levels are not sufficiently elevated, so her first sign of puberty usually appears as bumps or swellings beneath her nipples. Often, one breast will begin to develop before the other or will end up slightly larger than its partner; this is very common. Her vagina and uterus start to mature and she may notice an increased vaginal discharge; sex hormones subtly sculpt her torso to redistribute fat around her buttocks, widen her hips, and slim down her waist, a transformation engineered to accommodate a future fetus and undoubtedly to attract a male suitor when she is older. Within a few months of initial breast growth, sparse pubic hair sprouts on her vulva. She grows taller rapidly, so much so that her joints ache from time to time with "growing pains." Hormonal activity may result in oily skin, especially on her face, setting the stage for pimples or full-blown acne.

All of these changes typically precede a girl's first menstrual period (menarche), which on average occurs when she is between 12 and 13 years old and weighs about 105 pounds (about 47 kilograms). The onset of menstruation usually means she is able to become pregnant—but not necessarily. Occasionally, hormone levels may be high enough to cause her uterine lining to thicken each month and be expelled as menstrual fluid, but not high enough to ripen ovarian follicles. Until then, she remains infertile.

Right before and during the first day or so of her menstrual periods, she may suffer from pelvic cramping due to hormonelike chemicals called *prostaglandins* that trigger uterine contractions. In most cases, these cramps are mild and can be alleviated with heating pads or a massage, over-the-counter nonsteroidal anti-inflammatory medications, or some forms of exercise; painfully debilitating cramps may have serious causes and require medical attention.

Girls continue to grow taller and gain weight, albeit more slowly, until they are age 15 or 16; by that time their genitalia has matured, and their full complement of pubic hair, underarm hair, and adult weight and height have been established. While it may distress girls in their mid-to-late teens to discover they perspire more heavily, it is a perfectly normal event caused by their maturing sweat glands producing the pheromones of sexual attraction.

MALES

In a 10- to 11-year-old boy, in response to increased testosterone, his scrotum darkens and his testicles enlarge, followed shortly thereafter by a burst in his overall growth. Like girls, he finds his skin is oilier and he may go through several bouts of acne. His height increases and, by the time he is age 13, accelerates, continuing even into his 20s.

When he is age 13 or 14, his penis lengthens and pubic hair develops. His hands and feet start to catch up with his increased height, his shoulders widen, and his chest deepens with pronounced muscular development. Some boys are dismayed to find their breast tissue developing (gynecomastia), but the condition usually reverses itself within a few months to a year. Facial hair generally makes its first appearance as a mustache, then later extends onto the face and jaws. Although a boy experiences erections from an early age, he is embarrassed that his now more noticeable erections occur spontaneously, some at the most inappropriate times. His voice, hoarse and cracked in response to testosterone prodding his vocal cords to lengthen, betrays his sexual transformation. As his larynx enlarges to accommodate his vocal cords, it protrudes from his neck as an Adam's apple. He notices that he perspires more, like his female counterpart, with accompanying body odor.

These developments are awkward for a teenage boy coping with the social and emotional complexities of awakening sexuality, especially when he compares himself to girls his age whose physical developments are a year or so ahead of his. By late puberty, however, when adult pattern pubic and facial hair are established and his penis has thickened in girth, he reaches his physical maturity in every respect except final height.

Emotional and Intellectual Development

Weathering the upheavals of puberty is a rite of passage for teenagers, and some are more successful at it than others. They enter as children, with children's perspectives, and are expected to emerge on the other side with an adult appreciation of their own sexuality along with the maturity to enjoy it responsibly.

In early adolescence, as they struggle with independence, teenagers careen from one extreme to another, from a desire to distance themselves from their family to a need for connection and reassurance; as time goes by and they begin to view their parents as less than perfect, they more strongly identify with others and rely more and more frequently on their peer group for validation, adopting the clothing styles and fads that are prevalent among their friends.

Out of a natural tendency to test their boundaries, teens may associate with troubled or dysfunctional companions and adopt destructive habits they carry into adulthood. Their emotions, ranging widely and with some frequency from euphoria to depression—largely in response to hormonal fluctuations—can make their healthy adjustment to outside pressures very difficult.

When they are 15 or 16 years old, girls usually reach a threshold in physical development, but emotionally both sexes are still immersed in identity issues that revolve around approval from friends and separation from family. Sports and other forms of competitiveness characterize this period, as adolescents struggle to find acceptable outlets for their energy and aggression. Although a teenager's sometimes uneven scholastic performance may reflect his or her emotional upheavals rather than intellectual difficulties, most older adolescents reveal deepening philosophical perspectives amid a greater capacity for abstract thought. This can be a time of rich discovery and opportunity for self-expression, manifested perhaps in artistic pursuits like writing and drawing, that reflect their inner transformations.

By the time they reach the end of adolescence, in their late teens, their emotional storms have subsided and they have developed a much more stable sense of themselves and their place in the world. Boys are more comfortable in their own skin, and girls have adapted to their mature bodies. Teens find their unique personalities emerging and discover they have integrated the lessons of their parents, functioning—in most cases—responsibly and comfortably in the absence of a guiding authority.

Sexuality and Gender Issues

As hormones burst through their bloodstreams, preteens develop an intense interest in sex and find themselves inexplicably attracted to someone who a year or so before had seemed beneath their notice. Younger teens veer between excruciating shyness and showing off, often relying on teasing in their early flirtations to express interest in one another. Unfamiliar emotions and urges have them confused and aroused, yet too embarrassed to seek information from adults. One of the ways they gain information is, once again, through masturbation, which at this age is driven almost entirely by libido and is directly associated with discovering their own patterns of sexual response.

As early as age 10 or 11, a **pubescent** boy may have erotic dreams and fantasize to a climax, although his orgasms are dry until he is a little older and his ejaculatory apparatus has matured. Dream imagery that results in nocturnal emissions, or ejaculations, represents the onset of "wet dreams." While boys this age can follow one orgasm with a second or a third, that capacity diminishes as they grow older so that, by middle age, most men need a refractory, or recovery, period ranging from a half-hour to a day before they can ejaculate a second time. The level of stimulation and arousal involved in a given sexual encounter significantly affects response, so that, even if a man is temporarily incapable of ejaculation, he may easily attain and maintain an erection.

Girls also masturbate, usually by clitoral stimulation or insertion of objects into their vaginas, both to learn about their bodies and to satisfy sexual urges. Some reports suggest that girls start to masturbate later than boys, and that much of their fantasizing focuses on the romantic undercurrents of a sexual encounter as much as it does on physical stimulation. Other evidence indicates that girls experiment with masturbation at about the same age as their male counterparts, they fantasize about overt sexual activity, and they experience orgasms while sleeping. This anecdotal data is obviously contradictory, but what is firmly established is that, although they are often slower than their male partners to reach orgasm, females are often **multiorgasmic** and may in fact need more than one orgasm to be completely satisfied. Females who do not masturbate could be unaware of their need for clitoral stimulation to have orgasms, so they engage in sex missing one of its most satisfying rewards, while males usually climax quickly from the thrusting of intercourse alone and are usually satisfied afterwards. It is important that teenagers understand these different response patterns because that knowledge, or its lack, can fundamentally affect their sexual compatibility with future partners.

Once teens start dating, interpersonal relationships rapidly take shape that can be intense. Overcome by "puppy love," many believe their passion for another is testimony to undying love, while others form frequent relationships they casually break, abandoning the other party to heartache. Older adolescents seem capable of more mature attachments manifested by a greater concern for another's well-being and less for their own infatuation. And some teenagers, for a variety of reasons, are simply not ready to engage in relationships of this sort and it is only until late adolescence or early adulthood that they emerge with the maturity and confidence to do so.

By now, the coping and behavioral patterns teens develop during puberty begin to crystallize, and so does their sexual orientation. Despite the difficulties that adolescence entails, teens who are certain of their heterosexuality have an easier time than those who struggle with homosexual leanings. In a society that encourages heterosexuality (and even that is encumbered with cultural and moral constraints that may restrict its expression in one way or another), being a teenager with homosexual urges can be extremely difficult. As if that were not enough, some children have anatomical abnormalities that become enormously significant at puberty; many of their developmental anomalies begin in the womb, when hormones first exert their influence on the gonads, and become compounded at puberty, when hormones accelerate sexual development. The genitals aren't the only organs affected; the embryonic brain is also imprinted by hormones, and it is unclear how the pubescent brain is affected by prenatal and postnatal forces in the determination of sexual leanings. It is well documented that the attitudes and feelings of both genders overlap in nonsexual areas of their lives: men's sensitivity that is said to arise from "being in touch with their female side" is a typical example. Sigmund Freud (1856–1939), the father of psychoanalysis, observed about a century ago that men and women combine each other's characteristics, and that neither is purely masculine nor purely feminine. This is still a viable theory driving countless studies in psychology workshops today that utilize tools like the Bem Androgyny Scale to measure the extent to which gender-related perceptions of males and females overlap. Because men and women share emotional attitudes and sensibilities, one would assume they might share sexual ones—as indeed they do. Transgenderism defines a broad category of individuals who identify to some degree with the opposite gender and includes transvestites, who dress in the clothes of the opposite sex but consider themselves heterosexual, and transsexuals, who undergo surgery to anatomically "become" the opposite sex so as to be physically in accord with what they feel themselves to be.

These brief examples hint at the diversity of sexuality and its subtleties. There is a great deal yet to be learned about why and how these differences arise, especially since many pseudopsychological analyses of earlier decades blaming overbearing mothers and weak fathers have been debunked. Scientific inquiry, which so far has focused primarily on the origins of homosexuality, may provide some answers.

HOMOSEXUALITY

The debate about whether sexual orientation, specifically homosexuality (and, by extension, bisexuality), has genetic or biological origins is ongoing. Regarded at various times and in varying degrees as "unnatural" (even if it turns out to arise from "natural" causes), homosexuality was once regarded even by the medical profession as deviant. But in 1973, the American Psychiatric Association's classic diagnostic and statistical manual dropped homosexuality as a mental disorder, and, in 1975, the American Psychological Association formally supported that decision.

A decade ago, great excitement accompanied a research announcement that a "gay gene" may have been found that would prove the genetic basis of homosexuality. The cultural implications of such a finding would have been profound in a world that frequently regards homosexuals as unworthy of the same protections and privileges as other members of society. Although evidence of the gay gene failed to materialize, there are signs that cultural attitudes are gradually changing. In 1999, for example, the state of Vermont legalized same-sex unions. Stopping short of allowing marriage licenses to be issued to a homosexual couple, the provisions of the law nevertheless ensure that gay couples enjoy equal "domestic partnership benefits." In a development that reflects the divisiveness surrounding this issue, however, Vermont's legislature is being pressured to weaken the bill. Nevertheless, several other states, such as New York, have passed sexual nondiscrimination acts and laws to protect homosexuals from abuse and discrimination. In addition, the Human Rights Campaign, the nation's largest gay and lesbian political group, is working with the states and the U.S. Congress to extend similar protections and the right to civil privileges to all gay, lesbian, bisexual, and transgendered people.

There is an expectation that, as experts find evolutionary and related evidence that sexual orientation has significant biological roots, taboos against its alternative expressions may continue to ease and be put to rest within a cultural perspective that is universally acceptable. The research results are intriguing. Studies comparing identical twins to fraternal twins reveal that if one identical twin is gay, the other has a 50 percent chance of being gay even if the twins are raised separately, while the fraternal twin of a gay person has a 20–30 percent chance of being gay. It seems clear that the discrepancy, if it results from something like two different placental chemistries in utero (as some propose it does), has a biological basis. The same causation is suggested in the "fraternal birth order effect" that is named for this puzzling phenomenon: for each older brother a man has, the greater his odds of being homosexual. Other research into human sexuality supports animal studies that indicate certain brain structures have major influences on homosexual behavior. Yet other animal studies document widespread homosexual activity among their own kind, and thus pose the question: why would human evolution delete such pleasure-directed activity from the range of human sexual practices if it is so prevalent among his four-legged relatives? There appears to be less evidence linking nurturing and environment with homosexuality, but there is consensus that it plays a part, especially early in life.

Mainstream opinion holds, then, that sexual orientation arises from both prenatal and early developmental factors and, as a basic drive that is fully integrated by the time an individual reaches sexual maturity, cannot be changed at will, despite reports of some who have supposedly made the transition. Most experts interpret these "conversions" to mean that some individuals have successfully changed their behavior—for example, homosexual men marry women and have biological children with them—but they have not necessarily changed their predominant sexual orientation, and their ability to engage in heterosexual behavior merely suggests they can tap into a broader spectrum of sexuality than others might.

PRECOCIOUS PUBERTY

Precocious puberty is the development of sexual characteristics in girls younger than age 8 and boys younger than age 9. Often there is no obvious cause, but the rising cases of precocious puberty in Western countries during the last few decades have led to speculation that, in addition to increasing obesity among American children, growth hormones in animal feed, pesticides, and environmental pollutants are responsible. Known causes of the condition are growths on the adrenal glands or gonads, and central nervous system disorders. But any case of precocious puberty, no matter how benign, should be thoroughly evaluated by a medical professional because premature development needs to be arrested or even reversed; if not, affected children may not reach their full height. Another danger is psychological: if puberty is difficult for teenagers, it is that much more difficult for a younger child who may be the only person in his or her age group exhibiting any of puberty's characteristic physical changes.

DELAYED PUBERTY

Often characterized by short stature, delayed puberty is most frequently identified in girls over age 12 who have exhibited no breast development and in boys over age 14 who have exhibited no testicular development. Another significant diagnostic signpost is a lapse of five years between two major developments in secondary sexual characteristics—between breast development and the onset of menstruation in girls, for example. Like precocious puberty, it may have no discernible cause or it may be due to hormonal irregularities that can be corrected. Malnutrition, anorexia nervosa, and chronic infections may play a part. In other cases, chromosomal abnormalities that affect gonad development are implicated. Just as in precocious puberty, thorough medical evaluation is recommended so that appropriate treatment can be started as soon as possible, and caregivers need to be alert to the psychological implications the syndrome might have for young people.

Most cases of abnormal puberty are due to biochemical problems and can be successfully treated if the children exhibiting symptoms are comprehensively examined and their condition diagnosed early. Other developmental problems are much more serious.

ABNORMAL AND AMBIGUOUS GENITALIA

Many babies are born with abnormally functioning or malformed sexual organs (gonadal dysgenesis or aplasia), and some arrive with ambiguous genitalia that make it difficult to distinguish male from female. This latter group, clinically known as intersexed (see "Terms Related to Sexuality"), is at great risk, not from a physical standpoint but from a psychosexual one arising from the wrong gender assignment. An example is a child whose genetic and cerebral sexual differentiation make her a female but who is raised as a boy because her genitals more closely resemble a male's (see "Genital Ambiguity, the Dominican Republic, and Intersexuality"). At puberty, or even earlier, when her genetic identity emerges, she may develop serious psychological problems related to cultural gender roles and sexual issues that are, by then, resistant to treatment. For this reason, obstetricians and other neonatal professionals should be specially trained to determine the true gender of babies born with ambiguous genitalia.

The most common sexual anomalies in newborns, most of which arise from chromosomal abnormalities and can vary markedly in degree, are:

- Hermaphroditism—sexual organs combine male and female genitalia in various stages of development.

Genital Ambiguity, the Dominican Republic, and Intersexuality

The psychological trauma associated with ambiguous genitalia in many cultures does not seem to be an issue in the Dominican Republic where a common genetic disorder produces a male intersex population known as the "guevedoces" (also known locally as the "machihembras," which translates to "first women, then men").

In the womb, after a soon-to-be-male embryo's Wolffian ducts become internal gonads, his testes produce testosterone. But among guevedoces, a gene mutation interferes with their ability to convert testosterone into another kind of androgen, dihydrotestosterone (DHT), that is needed for the development of male external genitalia, so they are born with every appearance of being female. Local physicians are adept at identifying newborn guevedoces, so there are no surprises when what looked like a female turns into a male after androgen surges reverse the condition at puberty and the genitals mature along normal lines. Because their society is so accepting and their transition through adolescence relatively uneventful, guevedoces serve as fascinating case studies of how the attitudes of a tolerant society avert the serious psychological repercussions such children suffer in other cultures.

"Guevedoces" means "penis at twelve" and refers to the child's age at puberty when "his" clitoris develops into a man's penis and the labia swell preparatory to becoming scrotum into which the descending testicles drop. These males have the normal chromosome complement and, except for thin beards and smaller prostates due to embryonic androgen deprivation, they mature into adulthood as fully functioning sexual males capable of fathering children who, if male, could also inherit the condition and be born as guevedoces themselves.

- Hypospadia—refers to malformation and relocation of the male urinary passage or organs, although the infant is unquestionably identifiable as male.
- Cryptorchidism—one or both testicles are undescended.
- Klinefelter's syndrome—caused in males by an extra X chromosome (or more), XXY instead of XY, so there are forty-seven (or more) chromosomes present in the cell instead of forty-six. It leaves men sterile and, although symptoms vary depending on the number of extra X chromosomes, results in some breast development, small gonads, and lowered testosterone production.
- Turner's syndrome—a condition in females caused by the absence of an X chromosome; they are X0 instead of XX. Girls are sterile, but hormone therapy with oocyte donation or embryo transplant may allow a successful pregnancy.

Many of these genetic disorders involve improper hormone production during critical stages of embryogenesis. One such condition is congenital adrenal hyperplasia (CAH) that causes elevated secretions of androgens during embryogenesis and produces masculinized genitalia in newborn girls. It can be detected at birth with a blood test and should be treated with hormone therapy throughout life to develop and maintain her feminine traits. Other conditions can be treated surgically to restore normal appearance to genitalia, but often the patient is left infertile or may require hormone supplementation to develop secondary sexual characteristics and to function sexually as an adult.

AGING

Women, Menopause, and Sex

Unlike men, whose sperm production continues throughout their lives, women are born with a fixed supply of oocytes that are depleted over time. As Chapter 1 pointed out, stored oocytes are suspended from further development until ovulation and fertilization. During a woman's reproductive years, new egg follicles are always in the process of maturing but, because only one releases its ovum each month, many others must die. This is why, by the time a woman is 40 years old, her egg supply is thought to be only about 2 percent of what it was at her birth. In the Middle Ages, women who survived childbirth often died of other causes by the time they were little over 50; thus their fertility lasted about as long as they did. Nowadays, with humans routinely living into their eighth and ninth decades, many women experience as many years of menopause as they do of fertility. Recent research challenges whether this will continue to be the case. Studies find that, at least in mice, egg production may continue throughout the animals' lives. If this proves true in humans, the implications for treating infertility and the hormonal disruptions associated with aging could be enormous.

The mechanism of negative feedback discussed in Chapter 2 is well illustrated by menopause, also known as the climacteric. When hormone production from the ovaries slows, the hypothalamus reacts by prompting the ovaries to produce more. Many believe these hormonal surges to the "hot flashes" most women experience during menopause, the brief but intense periods of warmth and redness that suffuse the body, especially the upper torso and face, and often produce copious perspiration. Although uncomfortable or socially embarrassing, hot flashes are mostly notorious for interrupting or disturbing sleep.

Mammals, other than humans, rarely live long enough to undergo menopause; their life span and reproductive capacity generally coincide. It is a bit of a puzzle that humans outlive their reproductive years, because evolution, which highly values fertility as a measure of fitness for survival, does not tend to support nonreproductive longevity. One theory proposes that evolution hasn't had time to catch up with the rapidly increasing life span of the human race. Another is that human reproductive function actually does continue past menopause, in a sense, through grandmotherly nurturing; that is, humans' large brains have created complex social structures in which the long-term survival of mothers as grandmothers enhances the survival of succeeding generations. Others reject this "grandmother hypothesis" entirely, declaring that menopause is simply an adaptive mechanism that prevents women of advancing age from bearing offspring when they are unlikely to survive the child's dependent years.

Another theory proposes that the timing of human oogenesis and depletion, given the quantity of oocytes that are produced, ensures that only the highest quality ova survive to become suitable candidates for fertilization. This hypothesis has some basis in theories of aging, or senescence, which suggest that the advantages of an early-in-life trait outweigh, or at least equal, the disadvantages of a later-in-life trait: in this case, fetal development of numerous high-quality oocytes outweighs the evolutionary disadvantage of fertility's cessation. A woman is let off the hook: because the quality of her ova is not dependent on her continued fertility, her life span can be extended at no cost to the species.

In human physiology, natural menopause—as opposed to induced or surgical menopause that results from the removal of ovaries—occurs, on average, around age 51, but it can range from as early as 39 (somewhat premature) to as late as 59 (somewhat late). It is by no means a sudden event; just as the signs of puberty appear as a result of slowly rising hormone levels, menopause occurs in response to slowly dwindling supplies of estrogen and progesterone. Technically, menopause is said to have occurred when menstrual bleeding has ceased (amenorrhea) for at least one full year. Experts recommend that women continue using birth control for that year, however, because residual hormonal activity has been known to produce "menopause babies."

Menopause may be said to start as early as ten to fifteen years before ovulation ceases, when a very gradual rise in pituitary hormones (FSH and LH) corresponds to the slowdown in ovarian follicular development. About four to five years before discernible ovarian function stops, when estrogen and progesterone levels drop low enough, women enter **perimenopause** and their menstrual periods become irregular. Common symptoms include hot flashes; headaches, sometimes severe; insomnia; vaginal dryness; forgetfulness or temporary inability to concentrate; and aching joints. Depression or sudden anger can be a direct result of hormone imbalances, much as they are in puberty, and occur most often during perimenopause, while other emotional difficulties that crop up after menopause may have more to do with a woman's self-image as a sexual being and contributory member of the society her culture embraces.

Obviously, then, a woman's reproductive life ends at menopause. Certain physical changes in her body, such as thinning of genital tissue and vaginal dryness, can make sex uncomfortable, but these are easily remedied with the help of lubricants or other vaginal preparations commonly found in pharmacies. Some women experience a temporary lessening of libido, others experience a permanent decrease, and some experience no change at all or, in fact, report a heightened pleasure from sex due to external factors like more privacy (if her children are grown and living apart from her) and the liberating effect of natural, worry-free contraception.

Hormone replacement therapy—the practice of prescribing estrogen and progesterone supplements to menopausal women to substitute for those nature no longer produces—became popular during the 1990s as the baby boom generation began to age. Not only was HRT, as it is commonly known, supposed to help prevent cardiovascular disease, it would slow or halt osteoporosis, reduce the side effects of menopause, help preserve libido, and provide general benefits like keeping an aging woman's skin and other tissues more supple. The downside of HRT was an increased chance of breast and uterine cancer, but, particularly for those with cardiovascular risk factors, the benefits were felt to outweigh the risks. In 2002, however, the National Institutes of Health abruptly cancelled the Women's Health Initiative, a study that evaluated the most popular formulation of HRT, because early findings did not support cardiovascular benefits after all and established that the risk of breast cancer was slightly higher than had previously been predicted. In 2003, further studies determined that HRT increased the risk of both breast cancer and heart disease. Although hormone replacement helps prevent osteoporosis, there are other drugs that adequately treat this chronic condition, so the medical profession, for the most part, now recommends women use HRT only for a short time and only to relieve the most severe symptoms of perimenopause and menopause. Other hormone supplements such as **topical** testosterone creams are still being prescribed for menopausal women suffering from sluggish libidos.

Men, Andropause, and Sex

For a long time, few believed that a so-called male menopause existed, especially since men do not have the overt manifestation of hormonal depletion that women do when menstruation ceases. The assumption has therefore been that, in the absence of disease, men suffer from few reproductive or sexual difficulties as a result of aging.

In recent years, however, there have been increasing reports that many men do experience erratic and lowered hormone levels, specifically testosterone, that lead to decreased sexual function and several other symptoms. Men are seeking treatment in ever greater numbers, partly because of publicity surrounding the relief menopausal women have obtained from hormone treatment, but also because today's middle-aged population seems determined to postpone the effects of aging for as long as possible. Increasingly recognized as "andropause," male menopause causes symptoms in about 25 percent of men in their 50s and 50 percent of men in their 60s. Besides sexual dysfunction—reduced ejaculatory trajectory and lowered sperm count, diminished sexual interest, and difficulty achieving or maintaining an erection—the other major symptoms men take to their physicians are decreased energy and muscular strength and depression. Medications containing testosterone to treat these symptoms and drugs to restore blood flow for men suffering from **erectile dysfunction** are increasingly available.

Like HRT for women, testosterone replacement for men is controversial. Available by injection in a doctor's office, through a skin patch, or from a gel, it can have very serious side effects, such as a heightened risk of heart attack and stroke and, alarmingly, an increase in the growth rate of already-present prostatic tumors. Yet many feel it is worth the risk because restored erectile function and libido, increased muscle tone, and elevated mood have made those who suffer from andropause feel better—and younger.

People are increasingly reluctant to accept a diminished sex life as a condition of growing older. But just as their sexuality is shaped prenatally and postnatally by a combination of forces, not just hormones, it will likely take more than hormones to restore the urges and stamina of youth.

SUMMARY

Sexuality is not a start-and-stop response to specific stimuli, but a combination of forces interwoven by biology and culture that finds outlets along a broad spectrum of behavior. The cumulative effects of fetal imprinting, biochemistry, and environmental inputs emerge at puberty to shape sexuality and set the pattern of its expression. A relatively constant drive during the reproductive years, it is interrupted at middle age by hormonal upheavals that may blunt its demands and sometimes its rewards. From the innocent displays of sexuality revealed by young children in everyday play to the libido and virility boosters that older people seek, one way or another, subtly or not, sexuality exerts its influence throughout all of human life.

5

Myths and Early Discoveries

Anyone pondering the sex life of Stone Age people might be forgiven for envisioning rather brutish, furtive copulation in the shadowy recesses of caves. But if archaeological evidence pointing to a more sophisticated sexual culture is valid, ancestral humans created erotic sculptures, painted sexually graphic scenes on cave walls, distilled aphrodisiacs from herbs to heighten their sensual enjoyment, and participated in prostitution and homosexual behavior. The artifacts make it clear that prehistoric people knew quite a bit about sexual pleasure, but there are few clues to reveal how well they understood reproductive physiology.

On the other hand, written records say a lot about what their descendants did and did not know. For centuries, from ancient Greece to medieval Europe, people stubbornly supported preformation, the belief that a tiny, fully formed person existed inside each sex cell. Spermists suggested it resided in sperm, ovists maintained it was found in ova, and both groups thought it began to grow when fertilization occurred. It wasn't until after the Middle Ages and the invention of the microscope that scientists began to abandon preformation in favor of epigenesis, which holds that embryos undergo stages of differentiation and development. The preformation hypothesis was finally discarded early in the nineteenth century.

While the understanding of human biology grew steadily from the seventeenth century on, it was the introduction of Darwinian evolutionary theory and Mendelian genetics in the latter half of the nineteenth century that transformed science and led to deciphering the structure of DNA in 1953. But there was a downside: learning how traits were inherited gave rise to socio-political movements known as social Darwinism and eugenics. Supporters experimented with selective breeding programs intended to improve society by encouraging or discouraging, respectively, reproduction among the "fit" or the "unfit." Horribly exploited at the hands of the Nazis during World War II, these movements were abandoned for a time, only to be resurrected in another form late in the twentieth century.

EARLY RECORDED HISTORY

Two thousand years after the ancient Egyptians developed a pregnancy test that is thought to have been highly accurate, the Greek physician Hippocrates (460–ca. 410 BCE) suggested that semen and menstrual fluid contained the "essence" of the male and female that was passed on to their offspring. This essence included acquired characteristics, changes in the parents produced by environmental influences—a hearing impairment caused by excessive noise, perhaps. The philosopher Aristotle (384–322 BCE), a student of Hippocrates who rejected preformation and his teacher's belief in the heritability of such characteristics, proposed that sperm tempered or "cooked" menstrual blood to give substance and sustenance to the embryo, a belief further supported by the absence of menstrual flow during pregnancy. Although Aristotle amassed compelling data to prove that traits acquired by parents could not be inherited, the misperception lingered until very late in the nineteenth century when the mechanics of genetic inheritance were clarified.

The work of the ancient Greeks intrigued Islamic scholars who interpreted and expanded on that knowledge in texts that were later translated from Arabic and incorporated into the medical curricula of medieval Europeans. For many centuries, the teachings of the Greek physician Galen (129–ca. 210 CE) formed the basis of scientific and medical culture. His descriptions of human circulatory and nervous systems and his observations about reproductive physiology dominated medicine for over 1,000 years. A prolific writer, the anatomist was one of history's most influential physicians, but his beliefs contributed to the "tyranny of Galen"—collective misinformation so thoroughly integrated into medical thinking that for centuries it diverted scientists from pursuing more fruitful avenues of investigation.

During the Middle Ages (ca. 475–1450), many supposed that reproduction depended on the existence of female sperm, and some anatomists thought the ovaries, as **analogs** of the testicles, produced a spermlike substance. Determination of the embryo's gender was believed by some to rest on the resolution of the conflict between male sperm and female sperm (or other female "matter"); whichever was the stronger was more likely to transmit its own gender to the embryo. If the male seed encountered female seed of equal strength, neither gender could dominate, leaving the embryo to develop as a hermaphrodite. Another widespread misconception was that the womb was divided in half; if the male seed "landed" on the right side of the uterus, the child would be male; if left, a female.

Other medieval scientists supported pangenesis, the idea first proposed by Hippocrates that semen contained small representations of the body's organs and appendages that would later give rise to their counterparts in the embryo. This contrasted with preformation's miniature person (homunculus) already poised for growth in sperm or ova. Preformation was buttressed by distinguished scientists who, peering through crude, early versions of the yet-to-be-invented microscope, claimed to see these tiny persons in male sperm and published detailed drawings of them in prestigious medical journals. Despite preformation's wide acceptance, it presented a troublesome dilemma that was, of course, never resolved: how each homunculus could contain within itself a tiny being that, in turn, contained a tiny being, and so on.

Deeply embedded in science at the time was the influence of religion. A popular explanation for creation and man's place in the universe was the Great Chain of Being, which claimed that the Earth was only about 6,000 years old, that all living things had been created at that time, and that life forms had remained unchanged ever since. The least complex or least perfected life forms (rocks and plants) were at the bottom of the chain and, naturally, a human (male) sat at the top. Only the angels and God were depicted as superior to mankind. Related to this vision was a conviction that the soul, as the giver of life and therefore the catalyst of reproduction, was the "breath" of the body and inseparable from it. In the 1600s, however, French mathematician and philosopher René Descartes (1596–1650) suggested that since the mind operated independently of the body, the two were separate. This mind-body dualism was a major shift away from the old Aristotelian view of soul-body unity and marked a departure from a central theological tenet of the time. Thus Descartes' philosophy helped usher in the eighteenth century's Age of Enlightenment, when experimentation and objective analysis, rather than religious doctrine, underlay scientific theory. It is worth noting that the medievalists' view that mind and body were inextricably linked was not naive; dormant for centuries, a renewed interest in the mind-body connection is evident in modern approaches to healing.

By time the Renaissance had spread throughout sixteenth-century Europe, Gabriello Fallopio (1523–1562) had described the Fallopian tubes and rejected the proposal that ovaries were analogous to testicles; Hieronymus Fabricius (1537–1619) had determined that animals developed from eggs; and Marcello Malpighi (1628–1694) had used the newly invented microscope to study the embryonic development of chickens (yet he continued to support preformation). These individuals, and many like them, made historic contributions to natural science. Their work also dispelled some older misconceptions like spontaneous generation, which embodied the belief that life could arise from **inorganic** matter. William Harvey (1578–1657), whose work on the circulatory system overturned fourteen centuries of Galen's teachings, made especially significant contributions to embryology. A student of Fabricius, Harvey was the first to propose that both the mother and the father contributed equally to the creation of an embryo, which then underwent the epigenic stages of differentiation and development; he also described the early embryonic structure known as the **blastoderm**. And in 1667, Walter Needham (1631–1691) revealed that the umbilical cord was not an organ of respiration, as had been thought, but of nutrition.

Despite these advances, there was still much to be learned—and unlearned. Fabricius d'Acquapendente (1612–1667), an Italian anatomist, performed valuable early studies in embryology, but misidentified the role of sperm in fertilization. Others, such as Anton van Leeuwenhoek (1632–1723), who confirmed the existence of sperm cells and added much to the understanding of bacteria, insisted that he had observed homunculi inside sperm he examined under the microscope. Nevertheless, by the middle of the 1700s, preformation was increasingly challenged by data gleaned from studies in chicken embryology and Caspar Wolff's (1738–1794) description of embryonic germ layers. Wolff also suggested early principles of biogenic law, observing that all animals look very similar early in embryonic development, diverging into recognizably different species only as they continue to develop in utero. Preformation finally gave way to this growing understanding of embryogenesis.

Carolus Linnaeus (1707–1778) bridged, in a sense, the old world view of living things as a hierarchy of organisms with the most specialized class at the top, and the current world view, which categorizes organisms into groups from species through kingdoms. This Swedish biologist created the Latin classification system known as taxonomy; with modern additions of subcategories that reflect the greater distinctions now observed within species, Linnaeus' system is still used today.

THE AGE OF ENLIGHTENMENT

In the 1800s, as embryology was beginning to emerge as a separate discipline, Karl Ernst von Baer (1792–1876), who verified the development of germ layers, confirmed the existence of female ova. In 1842, German embryologist Robert Remak (1815–1865) further identified the germ layers as the mesoderm, ectoderm, and endoderm. Pangenesis, Hippocrates' old premise that various parts of the body contributed portions of themselves to the embryo, continued to be widely accepted, particularly after Charles Darwin (1809–1882) reinforced it with his description of "gemmules." The young British naturalist, who was soon to revolutionize the study of human origins, suggested that these tiny particles in each organ migrated to the gonads and transmitted acquired changes to the corresponding organ in the offspring.

Concurrent with the eighteenth and nineteenth centuries' advancements in scientific reasoning and the lessening of the Church's influence was increasing skepticism about the Great Chain of Being. French zoologist Georges-Louis Leclerc Buffon (1707–1788) not only suggested that life forms changed over time but went so far as to assert that the earth was at least 75,000 years old. His claim that species evolved was supported by Erasmus Darwin (1731–1802), Charles' grandfather and a physician who for some time had been persuaded that humans were products of evolution. Jean-Baptiste Lamarck (1744–1829), the elder Darwin's contemporary who was instrumental in establishing biology as a distinct science, made notable contributions to evolutionary theory. His theses, embedded in Lamarckism, were also founded on a belief in acquired characteristics; he wrote that environmental pressures created a need in an animal that resulted in structures evolving to meet that need, and these structures or adaptations were inherited by offspring. He used the antelope as an example: predators chasing it (an environmental pressure created a need for escape) gave rise to stronger legs (structures arose to meet the need) that, once developed (evolved), would be inherited.

It wasn't long before Lamarckism was supplanted by Charles Darwin's theories of evolution, which spelled out how species evolved from simpler organisms and achieved diversity through **natural selection**. German biologist Ernst Haeckel (1834–1919), Darwin's immediate and enthusiastic supporter, developed his "ontogeny recapitulates phylogeny" doctrine based on Darwin's work. It states that human embryogenesis (ontogeny) mirrors the evolutionary history (phylogeny) of animal species; some aspects of the doctrine are still accepted today.

DARWINIAN EVOLUTION

Charles Darwin was by no means the first to propose the concept of evolution, but he was the first to document how it occurred. His mentor, Charles Lyell, had earlier noted anatomical changes during embryogenesis that suggested the common origin of different species. Thomas Malthus' (1766–1834) observations regarding the expansion of populations beyond the ability of resources to sustain them helped form the basis of natural selection, a major thesis that Darwin and his collaborator, Alfred Russel Wallace (1823–1913), introduced jointly in 1858.

Ernst Mayr, one of the most eminent zoologists in the world today, calls Darwin's work "the greatest intellectual revolution experienced by mankind." As Chapter 6 will show, the knowledge accumulated over the last century in reproductive science, especially developmental embryology and genetics, is so intimately linked to evolutionary processes that a comprehensive study of the former requires a basic understanding of the latter.

The core of Darwinism, in very simple terms, is this: **adaptation** and natural selection render an individual more "fit"; because of his greater fitness ("survival of the fittest" or "principal of preservation," as it is also called), he is more likely to reproduce than a less fit individual because he survives, and survives for a longer period of time; thus the less fit are gradually removed from the population. The more fit have been naturally selected and the entire population has changed—it evolved. Continued variation, combined with **genetic drift** and the migration of species, will finally cause populations within a species to break off and become new species (speciation).

In more general terms, then, Darwin proposed that new species arise from an original one through branching descent with modification, a term that means all forms of life arose from a common ancestor (as proven by the fossil record); that evolution yields diversity in organisms and species through natural selection; that inheritance of favorable variations causes the gradual evolution of whole populations; and that species do not remain constant. Although sophisticated genetic and other analyses have confirmed that evolution indeed occurs, there is still debate among modern biologists about how natural selection causes change in species (Chapter 6 discusses the most salient of the differing opinions).

There are two common misconceptions about natural selection. One is that it "improves" the species; this is not necessarily the case. Evolution does not have an agenda; it is a natural process by which organisms adapt to existing conditions in their environment; thus any improvement seen in the capability of one species to survive in a given environment may not be an improvement for that same species in a different environment. The other misconception is that selection occurs by random accident or chance and therefore cannot lead to greater fitness (and the increased opportunity for reproduction offered by longevity and survival). The only part of natural selection that is completely random is mutation or gene reshuffling during the crossing-over phase of meiosis. The other part, sexual selection, is not random. It occurs all the time in nature, as when rams battle fiercely for dominance in head-butting competitions. The victor establishes his right to mate with the herd and pass on his genes, genes that contribute to strength and courage; these qualities favor survival of the species. Thus sexual selection occurs throughout the animal kingdom, not just among humans (see Chapter 2). It may be instinctive, unconscious, or culturally influenced (or all three), but it is not random. Finally, natural selection is not random in its result—the inevitable progression of evolutionary development from the simple to the complex, from a common ancestor to the variety and complexity of species on earth today. It was natural selection, rather than the concept of evolution formerly articulated by his predecessors, that was Darwin's great idea. It was flawed only by his conviction that acquired characteristics could be inherited.

Although mutations are a fundamental concept in evolution—and, of course, in genetics—Darwin himself did not define them. That fell to Hugo de Vries (1848–1935), a Dutch botanist who coined the word *mutation* in 1901 to signify a change in genes. Some mutations or genetic reshufflings that occur in crossing over are beneficial if they result in desirable characteristics; an example among hunting tribes of people might be increased height or improved eyesight. Others, such as those causing disease, are harmful. If the mutation affects the sex cells, it is a germline mutation and is inherited; if excess radiation exposure were to damage the DNA in a person's sperm or ova, the damage would pass on to the offspring and cause genetic defects. If the mutation affects the body cells, it is a somatic mutation and is not inherited; if tobacco smoke damages a person's lung tissue so severely that cancer develops, the offspring will not be born with damaged lungs or cancer.

In 1871, Darwin published *The Descent of Man and Selection in Relation to Sex* in which he examined human evolutionary history. Suggesting that man was not created in God's image, he stated, in some of the book's concluding remarks, that "Man still bears in his bodily frame the indelible stamp of his lowly origin."

MENDELIAN GENETICS

About the same time that Darwin was rewriting human history, an Austrian monk named Gregor Mendel (1823–1884), the father of modern genetics, was conducting crossbreeding experiments during which he identified the units of heredity. His contribution to genetics cannot be overstated, but unfortunately his work was overshadowed by Darwin's and was not identified as the critical missing piece of evolutionary theory until early in the 1900s, when Hugo de Vries and other scientists tested and validated it.

A mathematician and scientist who tended the monastery's gardens, Mendel was curious to learn why his plants displayed atypical characteristics from time to time. During his experiments, which he carried on for several years, he observed that certain "factors" (the term "gene" had not yet been coined, and did not originate with Mendel) seemed to influence the characteristics of his plants. Drawing on his education in mathematics, Mendel applied statistical analyses to establish inheritance patterns, the concept of dominant and recessive factors (or genes), and three laws of inheritance that underlie the science of genetics today. In so doing, he dispelled forever the old belief in blended inheritance, which held that each inherited trait represented a blending of both parents' traits. Like acquired characteristics, this mistaken perception had plagued Darwin, whose proposal that adaptive variations arise with succeeding generations could not be reconciled with variations that "blend" into oblivion over time.

As has already been explained, each person inherits one chromosome from his mother and one from his father. The corresponding genes—for eye color, for example—on each chromosome are called *alleles*. If one allele is for brown eyes (Br) and one is for blue (bl), they are heterozygous (Brbl). If both are for brown (BrBr), they are homozygous.

Mendel surmised that one allele may exert more influence in the offspring than the other; the stronger (Br) will therefore be the dominant allele and the weaker (bl), the recessive. If both alleles are for brown eyes (BrBr), they are homozygous dominant. If both are for blue eyes (blbl), they are homozygous recessive. The **Punnett square** in Figure 5.1 depicts the possible combinations that arise in the **genotype** of the offspring if maternal alleles (Brbl) combine with paternal alleles (Brbl). Notice that the homozygous dominant offspring (BrBr) will have brown eyes, as will the heterozygous offspring (Brbl).

Suppose the allele coming from one parent is faulty in some way? What are the chances the fault will affect the offspring? That depends on whether the allele is dominant or recessive (see Figures 5.2 and 5.3). Note in Figure 5.3 that if the defective allele is recessive, the offspring is a carrier who is not affected by the allele but who can pass it on.

Some characteristics that are transmitted through defective alleles are X-linked; that is, females are the carriers and males are affected. Hemophilia is a good example, and Figure 5.4 shows how it is passed on from mother to son. With a normal allele on one X chromosome that protects the mother from the defective allele on her other X chromosome, she is not in danger of having the disease herself but she can pass it on to 50 percent of her sons.

If the father is:	If the mother is: Br	bl
bl	Brbl BROWN (DOMINANT)	blbl BLUE (RECESSIVE)
Br	BrBr BROWN (DOMINANT)	Brbl BROWN (DOMINANT)

Br = Brown
bl = Blue

The **dominant allele** ("Br"), the allele that is expressed in preference to the **recessive allele** ("bl"). This is why an individual with both a dominant and a recessive allele will display the dominant characteristic.

Figure 5.1. Configurations of a normal genotype for eye color.

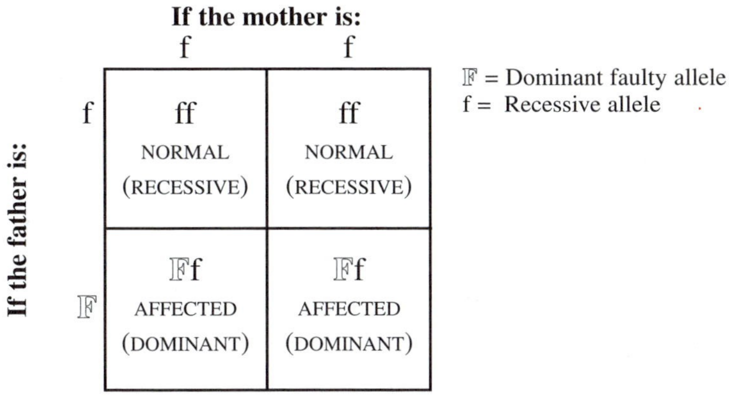

If the mother is:

If the father is:	f	f
f	ff NORMAL (RECESSIVE)	ff NORMAL (RECESSIVE)
𝔽	𝔽f AFFECTED (DOMINANT)	𝔽f AFFECTED (DOMINANT)

𝔽 = Dominant faulty allele
f = Recessive allele

Figure 5.2. Configurations of a genotype with a dominant faulty allele.

If the mother is:

If the father is:	F	𝕗
f	Ff NORMAL (DOMINANT)	𝕗f CARRIER (RECESSIVE)
F	FF NORMAL (DOMINANT)	F𝕗 CARRIER (DOMINANT)

F = Dominant allele
𝕗 = Recessive faulty allele
f = Recessive allele

The normal recessive allele (f) in the genotype shown in the upper right block at left will protect against the faulty one (𝕗), so the offspring, a carrier, will be unaffected.

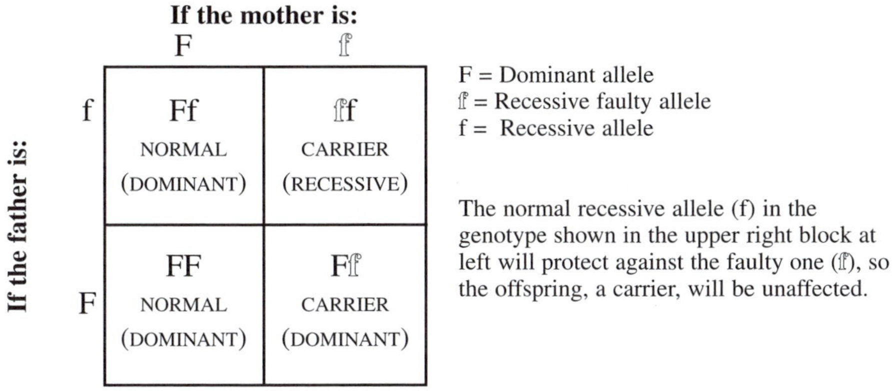

Figure 5.3. Configurations of a genotype with a recessive faulty allele.

Her daughters, like her, are protected from the disease, but 50 percent of them will become carriers. When one of the sons born with hemophilia has children, he will transmit the defective allele to his daughters and make them carriers, while his sons will be unaffected (see Figure 5.5).

Mendel conducted many other breeding experiments of great complexity, but just his results described here rank among the greatest scientific achievements in history. Dying in 1884, only two years after Darwin, Mendel did not live to see his work receive the wide acceptance it attained in the early 1900s, when it not only resolved the persistent misconception that acquired characteristics could be inherited, but illuminated Darwin's work on natural selection as well.

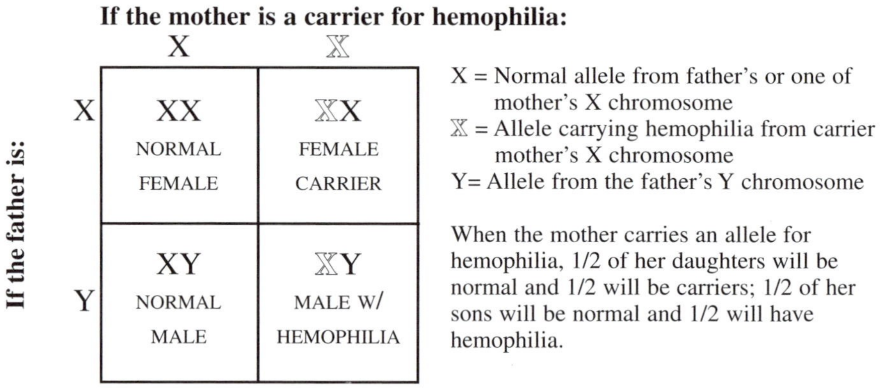

If the mother is a carrier for hemophilia:

If the father is:	X	𝕏
X	XX NORMAL FEMALE	𝕏X FEMALE CARRIER
Y	XY NORMAL MALE	𝕏Y MALE W/ HEMOPHILIA

X = Normal allele from father's or one of mother's X chromosome
𝕏 = Allele carrying hemophilia from carrier mother's X chromosome
Y= Allele from the father's Y chromosome

When the mother carries an allele for hemophilia, 1/2 of her daughters will be normal and 1/2 will be carriers; 1/2 of her sons will be normal and 1/2 will have hemophilia.

Figure 5.4. Configurations of a genotype with a sex-linked characteristic on the mother's X chromosome.

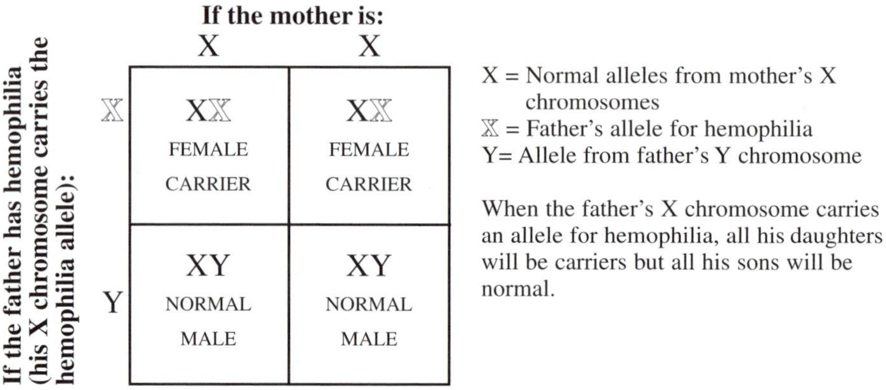

If the mother is:

If the father has hemophilia (his X chromosome carries the hemophilia allele):	X	X
𝕏	X𝕏 FEMALE CARRIER	X𝕏 FEMALE CARRIER
Y	XY NORMAL MALE	XY NORMAL MALE

X = Normal alleles from mother's X chromosomes
𝕏 = Father's allele for hemophilia
Y= Allele from father's Y chromosome

When the father's X chromosome carries an allele for hemophilia, all his daughters will be carriers but all his sons will be normal.

Figure 5.5. Configurations of a genotype with a sex-linked characteristic on the father's X chromosome.

DNA DECODED

In 1868, a Swiss scientist named Friedrich Miescher (1844–1895) became the first biologist to isolate nucleic acid in cells, but it wasn't until the early 1900s, shortly after Mendel's work became widely known, that the focus of investigation centered on chromosomes and their composition inside the cell nucleus. Edouard van Beneden (1846–1910) determined that different species have a set number of chromosomes and he identified the phenomenon of haploid chromosome formation in the cell. Scientists learned that chromosomes come in pairs and that X and Y chromosomes determine gen-

Social Darwinism and the Eugenics Movement

Late in the nineteenth century, the possibilities inherent in manipulating heredity triggered sociobiological movements that, for all their avowed humanism, had very negative consequences. British philosopher Herbert Spencer (1820–1903) speculated that the "survival of the fittest" premise—that the more robust (better adapted) individuals in a given environment are more likely to survive and reproduce, thus genetically passing on traits of greater fitness—could be applied to the races of mankind. Known as social Darwinism, Spencer's suggestion was that a given race whose members had certain weaknesses, perhaps a lower level of intelligence (by northwestern European standards) or evidence of questionable morality, rendered that population unfit. To avoid polluting the more fit race, the unfit must not be permitted to marry interracially. This, Spencer argued, was fully in accord with Darwinian natural law because it allowed the fittest to survive.

About the same time, Francis Galton (1822–1911), a cousin of Charles Darwin who was a respected anthropologist with an extensive background in agricultural breeding, approached the issue from a somewhat different perspective. His plan, known as eugenics (or "good genes"), was to improve human happiness and intelligence by reducing the birth rate of the unfit and elevating that of the fit by encouraging earlier marriages and providing better healthcare. Galton's proposals helped launch the eugenics movement, but, since Mendel's principles of heredity were not yet known, there was no way to effect the goals of the movement and so it languished for several years.

Early in the twentieth century, the eugenics movement reemerged when an American named Charles Davenport (1866–1944) proposed, rather expansively, to "improve the race" by breeding out of existence the "feeble minded." And who were they? According to Davenport, they were individuals whose behavior exceeded the bounds of polite society. Thus, promiscuous sexual behavior, thievery, assault, or family desertion fell within the realm of feeble-minded behavior. Soon the umbrella of undesirables included the poor and immigrants from southeastern Europe, who were judged to be more prone to criminality than immigrants from northwestern Europe.

Davenport proposed to sterilize poor people and to restrict immigration so that the entire population would be sanitized within 100 years. Shocking by today's standards, his overt discrimination was then viewed enthusiastically as the ideal cure for the ills of society. Most of the scientific community and prominent Americans such as Teddy Roosevelt supported it, and the few who expressed misgivings were soothed with repeated assurances that those who were sterilized would, of course, be treated gently and kindly (as befit a humane nation).

Collectively represented by the American Eugenics Society, the movement was ultimately able to put some brakes on immigration and even helped foster court decisions favoring involuntary sterilization. But as Adolf Hitler (1889–1945) rose to power in 1930's Germany and expressed his vast admiration for eugenics, its ominous implications began to emerge.

In World War II, Nazi enthusiasm for experiments based on eugenics carried out on concentration camp victims horrified the world and ultimately led to the rejection of the entire movement.

Buried for decades, the selective breeding practices that underlie eugenics are reemerging in a new form; the modern biotechnology that promises gene therapies to rid mankind of devastating inherited diseases also offers opportunities for tampering with the human genome to "enhance" the species. Creating designer babies who are genetically programmed to have good looks or greater intelligence or stronger muscles is becoming increasingly likely; already the genomes of human embryos can be prescreened for diseases. Adding or subtracting individual genes to customize the embryo is, many believe, a logical next step, but it is one that will have profound consequences for human culture (see Chapter 7 for much more on this topic).

---

der. Phoebus Levene (1869–1940) identified the nucleotides present in RNA and DNA. Others, led by Thomas Hunt Morgan (1866–1945), a pioneering geneticist who showed that genes are carried on chromosomes (and who, incidentally, was one of the few geneticists of his time who opposed eugenics; see "Social Darwinism and the Eugenics Movement"), explored sex-linked genes and the mechanism by which genes are exchanged during crossing over. Nevertheless, it wasn't until 1944, after years of incremental progress in deconstructing the cell's nucleus, that Oswald Avery (1877–1955) and his colleagues identified DNA as the molecule that carries genetic information.

With the announcement in 1953 that scientists had proposed a three-dimensional helical structure for DNA, history was made. Aided by information gleaned from crystallographic x-rays taken by Rosalind Franklin (1920–1958) and Maurice Wilkins (b. 1916), the team of Francis Crick (b. 1916) and James Watson (b. 1928) was able to construct a model of the DNA molecule within which, for the first time, it was possible to clarify the mechanics of reproductive replication and protein synthesis. In 1962, after Franklin's death four years earlier, Wilkins, Crick, and Watson were awarded the Nobel Prize for Physiology or Medicine for their stunning achievement.

SUMMARY

Adopting newly validated scientific principles about human reproductive biology—and discarding the invalid along the way—was a stumbling and occasionally stalled process, weighed down with centuries of misinformation and sometimes repressive religious dogma. But after the mid-nineteenth century, fueled by Charles Darwin's methodically reasoned theories of natural selection and evolution and Gregor Mendel's rigorously controlled genetic experiments, studies of human reproduction, especially at the molecular level, exploded.

Armed since the middle of the twentieth century with the chemical formula and structure of DNA, biologists have been unraveling the genome and divining the intricate processes of the cell at an ever-increasing pace. Chapter 6 continues the discussion of evolution and genetics from the perspective of the twenty-first century, as scientists continue to debate and refine new issues emerging from the momentous work of Darwin, Mendel, and many others.

6

Modern Perspectives: Reproductive Biology, Genetics, and Our Evolving Selves

By the late 1800s, disparate information about reproductive biology that had been collected over a period of centuries had been incorporated into rational models of human development by Charles Darwin and Gregor Mendel. In the 1930s, the new field of genetics merged with biochemistry and biophysics to become known more generally as molecular biology. In the meantime, scientists increasingly solved many of the mysteries surrounding reproductive physiology and biology that had puzzled them for centuries. Watson and Crick's publication of the structure of the DNA molecule in 1953, along with the 1959 discovery of the DNA polymerase enzyme that triggers DNA replication, opened up vast, exciting avenues of exploration and experimentation unprecedented in the history of life sciences.

This chapter reveals what biologists have learned in the last century about human reproduction and development. That knowledge has validated Darwin and Mendel and led to the birth of a modern subscience known colloquially as "evo-devo," or the evolution of development, founded on principles gleaned from reproductive biology, embryology, genetics, and evolution.

REPRODUCTIVE BIOLOGY

The end of the nineteenth century was marked by more than the revolutionary ideas of Darwin and Mendel. Earlier decades had witnessed two groundbreaking advances that had transformed childbirth from an agonizing and life-threatening ordeal into a less painful and less dangerous one: the use of anesthesia, and the introduction of sterile medical procedures that prevented the deadly streptococcal infection **puerperal fever** from being transferred to other obstetrical patients.

Although classical Roman and Arabian civilizations appear to have had significant obstetrical knowledge and even performed Caesarean sections, if only to save the fetus when the mother was dead or dying, most of their skills failed to transfer to the medical culture of Medieval Europe. This began to change during the Renaissance, partly because of the development of the printing press, which permitted more widespread dissemination of information. While Caesarean surgeries that would spare the life of the mother are reported to have taken place as early as 1500, it wasn't until antiseptic techniques were used in the late nineteenth century that the surgery became routine and was performed with the expectation that both mother and child would survive.

Before obstetrics became a formal medical discipline, midwives delivered babies in mothers' homes, where infection was not so widespread as it would later become in hospitals until sterile conditions prevailed. As the professional and economic threat that midwives posed to traditional male physicians began to mount, however, many midwives were accused of witchcraft and burned to death for practicing the healing arts. The increasing use of forceps and other surgical equipment further excluded them from practicing so that, by the nineteenth century, physician-dominated obstetrics ruled the childbirth wards. Nevertheless, midwifery continued throughout the twentieth century, particularly in rural areas of the United States, and today it enjoys a resurgence due in part to the medical establishment's recognition of its value (see Chapter 3).

Early efforts to control reproduction in the late nineteenth century and early twentieth century were manifest in experiments to artificially inseminate livestock and to introduce birth control techniques to the human population at large. The latter movement, focusing on barrier methods of contraception such as diaphragms, created friction and attracted controversy in medical and religious communities for a variety of economic, political, and moral reasons. Gradually, however, contraception gained worldwide acceptance, setting the stage for the arrival of the birth control pill in the middle of the twentieth century. Twenty years later, in the landmark U.S. Supreme Court case *Roe v. Wade*, American women gained legal approval to choose safe medical abortion as an alternative to unwanted pregnancy. At about the same time, abortion became legal throughout most of Europe.

Like abortion, treatments for infertility have been controversial, especially since 1978 when the first "test tube" baby was born in England. Lesley Brown was able to conceive and give birth to a daughter named Louise thanks to two pioneering physicians, Robert Edwards and Patrick Steptoe, who had perfected a method of in vitro fertilization (IVF). While their success was widely applauded by some, others objected on the grounds that the physicians were "playing God," and that artificially manipulating life was the ominous forerunner of a renewed eugenics movement. (See Chapters 7 and 10 for further discussions of the legal and ethical issues associated with assisted reproductive technology.)

A great deal of attention was given to embryology during the nineteenth century as well, particularly in terms of morphology, the form and structure of an organism. Toward this effort, the anatomist Wilhelm His (1831–1904) studied the formation of a series of **vertebrates** to determine the mechanisms of body formation. A subsequent breakthrough by Hans Spemann (1869–1941) that identified the "organizer," a group of cells that directs the development of the embryo, greatly illuminated ensuing studies like those of Viktor Hamburger (1900–2001). This pioneering biologist, much revered for his work in neuroembryology, relied on the Spemann organizer theory when developing his theses of programmed cell death and the role of chemical growth factors in the evolution of body tissues. Hamburger was also able to establish that the chicken embryo was a suitable model for embryologic experimentation, thus removing a major stumbling block posed by the lack of human embryos on which scientists could conduct experiments.

Before the first half of the twentieth century was over, biology was "modernized" when embryology, morphology, and a number of other disciplines were united under a common umbrella term known as the modern evolutionary synthesis. The emerging field of genetics was among them.

GENETICS

The Age of Genetic Engineering

As the previous chapter explained, Mendel's mathematics background gave him the tools to deduce the following three laws of inheritance:

- Law of Dominance. In a cross of two homozygous parents (both alleles of each parent are dominant or recessive) with contrasting traits (BrBr [in parent number one] × blbl [in parent number two], for example), only one of the traits will appear in the first generation of offspring, even if the offspring is a hybrid (Brbl). In other words, both Br and bl are independent traits and only one (brown eyes rather than blue eyes, for example; see Figure 6.1) will dominate in the phenotype.
- Law of Segregation. The two alleles responsible for a trait separate from one another during meiosis, so that each sperm or egg will have an allele for either a recessive or a dominant trait, but not both. In a cross of two heterozygous parents (each parent has a recessive and a dominant allele), therefore, one-fourth of the first generation offspring will display the recessive trait (see Figure 6.2).

If the mother is:

If the father is:	Br	Br
bl	Brbl BROWN (DOMINANT)	Brbl BROWN (DOMINANT)
bl	Brbl BROWN (DOMINANT)	Brbl BROWN (DOMINANT)

Br = Brown
bl = Blue

All offspring of homozygous parents display the dominant trait (brown eyes).

Figure 6.1. Law of dominance.

If the mother is:

If the father is:	Br	bl
Br	BrBr BROWN (DOMINANT)	Brbl BROWN (DOMINANT)
bl	Brbl BROWN (DOMINANT)	blbl BLUE (RECESSIVE)

Br = Brown Eyes
bl = Blue Eyes

Heterozygous alleles (Br and bl) separate during meiosis so that, when they recombine, one-fourth of the offspring will display the recessive (blbl) trait.

Figure 6.2. Law of segregation.

- Law of Independent Assortment. Alleles for one trait will be distributed to offspring independently of alleles for other traits (see Figure 6.3).

It is remarkable to reflect that Mendel formulated these principles over 100 years before genes and chromosomes were well understood. Verified during the first half of the twentieth century, his laws helped establish genetics as a separate science discipline. After James Watson and Francis Crick determined the structure of DNA, Crick's continued investigation into the subject led to his 1956 "central dogma" that declared genetic information goes in only one direction, from the genes that comprise DNA, to RNA, to proteins. About the same time, Severo Ochoa and Arthur Kornberg discovered DNA polymerase, the enzyme that drives DNA replication in the cell. This

If the father is:	If the mother is: BH	Bh	bH	bh
BH	BBHH BROWN TALL	BBHh BROWN TALL	BbHH BROWN TALL	BbHh BROWN TALL
Bh	BBHh BROWN TALL	BBhh BROWN SHORT	BbHh BROWN TALL	Bbhh BROWN SHORT
bH	BbHH BROWN TALL	BbHh BROWN TALL	bbHH BLUE TALL	bbHh BLUE TALL
bh	BbHh BROWN TALL	Bbhh BROWN SHORT	bbHh BLUE TALL	bbhh BLUE SHORT

B = Brown Eyes
b = Blue Eyes
H = Tall
h = Short

Heterozygous alleles (Br and bl, H and h) separate during meiosis so that, when they recombine, the assortment of the two traits (eye color and height) will appear in the phenotype independently of each other.

The results will always be the same:

- 9/16 of the offspring will show a dominant phenotype for both traits
- 3/16 will show dominant phenotype for the first trait and recessive for the second
- 3/16 will show a recessive phenotype for the first trait and dominant for the second
- 1/16 will show the recessive form of both traits.

Figure 6.3. Law of independent assortment.

breakthrough, which earned them the Nobel Prize in 1959, suggested that someday, outside of the cell, DNA could be duplicated for experimental purposes. In 1973, once Stanley Cohen and Herbert Boyer introduced a method of gene splicing (or DNA recombination) to remove a piece of DNA for insertion into another piece, scientists experimented with bacteria to determine if recombinant DNA would reproduce. When the altered bacteria successfully multiplied, recombinant technology was launched and genetic manipulation or "engineering" rapidly grew more sophisticated. Ethical concerns that began to arise about the potential misuse of the new technology in human subjects was addressed by National Institutes of Health cautionary guidelines; since the dire consequences some scientists had predicted never materialized, however, those particular guidelines have largely been retired.

In 1978, Stanford University successfully transplanted an animal gene, and in 1981 Ohio University produced the first **transgenic** animal into whose DNA some genetic material from another species was inserted. In 1985, the National Institutes of Health approved experimental gene therapy in humans in which a normal gene was introduced into a patient's DNA to treat disease that arose from its defective counterpart. Gene therapy is still experimental and still regarded as risky, but, as the next chapter shows, research is moving forward. A technique of the future could well be germline engineering, in which a gene substitution made in a parent's DNA—replacing a mutant harmful gene with a normal one—will be passed on to future generations, thus deleting the original mutation from the gene pool for all time. Genetic testing is already commonplace, allowing investigators to screen an individual's DNA to detect inherited mutations that predispose him to certain diseases.

Near the end of the twentieth century, a groundbreaking invention premiered that transformed the science. Kary Mullis, a California biochemist, introduced a **polymerase chain reaction (PCR)** technique that rapidly produced billions of copies of a given DNA molecule. Because large amounts of DNA are vital for experimental and applied genetics technologies, modern research would have been severely crippled without Mullis' Nobel Prize-winning contribution.

The Era of the Genome

Genetic engineering is a vibrant, fertile area of research made more so by the determination in 2003 that there are more than 30,000 genes in the human genome. While this figure may change as "gene-counting" technology advances, scientists now have the fundamental recipe for "tweaking" genes to alter cellular activity in the entire organism. Research on this front is expanding into disciplines like proteomics to map cellular pathways and determine how proteins direct the cells to operate the machinery of the body. Another area of investigation surrounds **transcription factors** and an **epigenetic code** that some scientists believe programs the cell to turn genes on or off. The cell directs that the proteins called *histones*, which are coiled within nuclear chromatin, undergo a chemical process known as methylation that seems to regulate on/off switches. Mastering the switches means that scientists can manipulate critical life processes.

The possibilities for and implications of genetic engineering are vast, affecting not only science but ethics, culture, and religion as well. With the new tools made available by genetics and proteomics, scientists are faced with staggering opportunity—and responsibility. These developments led, in the late 1990s, to President Clinton's establishment of the National Bioethics Advisory Commission to weigh the bioethical issues surrounding research on human cells and tissues. Although the Commission's charter has since expired, other panels have been convened to address the ethics of human embryonic stem cell research. It is extremely controversial because the critical experiments with stem cells that may result in cures for disease or the replacement of the body's organs also destroy the embryo. President George W. Bush, who opposed the research, directed in 2001 that federal funds could support research only on those stem cells already extracted and grown into cell lines in the laboratory. Because those cell lines are limited in quantity and genetic diversity, however, many scientists and others are seeking to overturn the directive. The issue will continue to be divisive as politics, science, religion, and culture collide. (Chapter 7 discusses embryonic stem cell technology and alternative means of developing genetic therapies in greater detail.)

Mutations and Genetic Variation

One of the insights gleaned from molecularly based research is that DNA is less stable than had originally been thought. While this does not change the orthodox view of genetics and heredity, it compounds the complexity of predicting just how genetic manipulation or mutation might affect gene expression. At its most basic level, a mutation is a change in one of the four bases comprising a segment of DNA (a point mutation), or changes in a specific segment of DNA. It may be due to damage from the external environment, such as exposure to radiation or certain chemicals like those used in chemotherapy for cancer; it may be because of DNA "mismatch," an error in replication; it may occur when DNA strands break; or it may be inherited. Ironically, despite the "fact" that acquired characteristics are not hereditary, a mutation acquired from the environment that affects the chromosomes of sex cells is an exception, because it can be passed on to offspring. Fortunately, the body can repair some of these mutations by mobilizing an array of proteins that joins broken ends, replicates missing pieces of a sequence, or even replaces damaged nucleotides. But if the repair mechanism is overwhelmed by repeated and persistent assault, such as years of exposure to tobacco smoke, disease can result.

Mutations can also occur as transposons, when sections of an entire chromosome are rearranged or when sections of one chromosome "jump" over to another, or they can result from a greater number of chromosomes in the cell than the normal forty-six (polyploidy). If a cell with two X chromosomes also has a Y chromosome (XXY), the embryo will have Klinefelter's syndrome. He will be a male despite two X chromosomes, simply because he has the Y chromosome carrying the transformative "male-making" SRY gene. This is why embryos with only one X chromosome, such as those with Turner's syndrome (X0), are females. The absence of a Y chromosome leaves their gender intact. So, while irregular X-Y chromosome arrangements can be relatively uncomplicated in terms of sex determination—if there is no Y chromosome, the zygote is female—they can lead to other complex problems (see Chapter 4).

This is not to say that all mutations are harmful. In fact, evolution depends on mutations, for they are the mechanisms by which individuals adapt and species vary (see, however, the later section, "Modern Theoretical Approaches to Evolution and Inheritance"). A mutation can be harmful or less beneficial in its initial manifestation, yet be perceived as more advantageous as natural selection deems the changes it produces to be desirable ones. Thus, in each case, a mutation's evaluation as harmful or beneficial should be made from the same perspective: its immediate impact on individuals or its long-term impact on species.

Chapter 5 introduced germline and somatic mutations in the context of their effects on cells during one's life, and explained that only when the chromosomes in the sex cells are affected by a mutation will it be inherited. If the mutation appears on the X chromosome (as occurs the most often) in ova or sperm, it is known as an X-linked mutation; if it appears on the Y chromosome (rarely) in sperm, it is a Y-linked mutation. If it appears on any of the other twenty-two chromosomes in the sex cell nucleus, it is an autosomal or somatic mutation. Thus four kinds of mutation are possible in Mendelian inheritance:

- autosomal recessive inheritance—both alleles must be mutants in order for the characteristic (phenotype) to be expressed;
- autosomal dominant inheritance—only one allele must be mutant for the phenotype to be expressed;
- sex-linked recessive inheritance—both alleles must be mutants; or
- sex-linked dominant inheritance—only one allele must be a mutant.

Mitochondrial Inheritance

Mitochondria are **organelles** in cellular cytoplasm inherited almost exclusively from one's mother. They are involved in energy production (they extract fuel by burning food) and contribute to hereditary characteristics because they have their own DNA (mitochondrial DNA, or mtDNA), albeit in much less quantity than that found in the cell nucleus. When cells divide in the body, the mitochondria separate independently and randomly into the daughter cells, so their effect on different cells in the body can vary. At fertilization, it is the maternal mitochondria that are retained in the larger body of the ovum, because the father's mitochondria are relegated to the tail of sperm. A daughter will pass her mitochondria on, but a son cannot because little, if any, of his mitochondria will be incorporated into his mate's ovum at fertilization.

MITOCHONDRIAL EVE

Because mitochondrial DNA is transmitted through the maternal line, genetic lineage on the mother's side can be traced. In recent years, investigators have found that the DNA of representative living Europeans linked them to one of seven genetic groups dating back several thousand years. Of even greater import was the finding that these seven groups most likely arose from a single group or a single woman that lived in Africa, known as Mitochondrial Eve. Scientists "discovered" her by tracking mtDNA mutations, which can be expressed in offspring, sometimes because a single point mutation disrupts the overall chemical pathways of the cell. Since mtDNA has no repair enzymes and cannot recombine, its mutations, unlike those of nuclear DNA, are more easily discerned.

It is important to understand the implications of mitochondrial inheritance and Eve. Mitochondria are only a part of the heredity pathway humans follow and, while their mitochondrial lineage may be easier to trace than that of nuclear DNA, it is nevertheless subject to variation along the way that can skew the evidence (for example, by the introduction of a little bit of male mitochondria into the ovum at fertilization). Furthermore, remember that Mitochondrial Eve is not the Biblical Eve or the original "mother" of everyone. She is simply the *most recent* common ancestor in the maternal line to whom living Europeans can be traced. There were other "Eves" before her; if the mtDNA of a population living 5,000 years ago could be traced, a more ancient Eve would no doubt be found. Moreover, if nuclear DNA could be traced back through the generations as easily as mtDNA is traced, it is just as likely a male would be found with the dubious distinction of being the most recent known common ancestor.

Therefore, mtDNA reveals only pieces of human heredity. The rate and type of mutations can skew timelines, the introduction of paternal mtDNA during fertilization can alter genetic material, and the influence of mtDNA on proteins synthesized by nuclear DNA obscures the record. Nevertheless, analysis of mtDNA is a unique tool that broadens science's overall perspectives on human development.

X-Inactivation and Genetic Imprinting

Although Mendelian genetics, like Darwinian evolution, has withstood 150 years of testing, scientists have discovered another route of genetic variation that, like mitochondrial inheritance, lies outside the model of the Punnett square. X-inactivation and genetic imprinting refer to differences in the way maternal or paternal genes are expressed. When a female is conceived, the zygote has two X chromosomes, one from its mother, and one from its father. A male zygote has only one X chromosome, from its mother, and a Y chromosome from its father carrying the SRY gene. The process of X-inactivation seeks to correct the overabundance of X-linked genes in every female. Very early in embryogenesis, an XIST gene on her X chromosomes encodes for a special kind of RNA that accumulates on each chromosome. Shortly thereafter, about the time cleavage begins, the RNA on one of those chromosomes begins to break down. The reasons for this are unclear, but the effect is to inactivate the entire chromosome, now known as a Barr body. Although the Barr body may have been the X chromosome coming from either the male or the female parent, it is specifically the *paternal* X chromosome that is kept activated in cells giving rise to the extraembryonic membranes (amnion, placenta, and umbilical cord). So, in these cells, the paternal X chromosome does not become the Barr body. Since females thus have two X chromosomes expressed in their bodies—some cells express the paternal X chromosome and some express the maternal X chromosome—they are genetic mosaics.

Genes (alleles) from both parents are usually expressed in cells, but in some cells only the alleles coming from a specific parent are expressed. This phenomenon is known as genetic imprinting, stamping onto chromosomes the "memory" of the parent from whom it came. Only the allele on the still-active chromosome—the allele with the memory imprinted on it—will be expressed, leaving the other silent. This kind of imprinting is not fully understood, although it is known that in most cases (but not all) genetic material from both parents is required for normal development to occur, which raises intriguing questions about the role of sperm RNA in fertilization and embryogenesis, discussed in the next section. Not surprisingly, imprinting's effect on the offspring varies depending on whose alleles are expressed (the mother's or the father's), and there is recent evidence that imprinting gone awry can result in disease. It also accounts for some of the tiny but real differences in identical siblings, each of whom expresses differently imprinted genes.

New Insights into RNA

Chapter 1 explored the more familiar mechanics of sexual reproduction, such as meiosis and fertilization, but technology has opened up new avenues to biologists who want to explore even more complex cellular processes. RNA, the cousin of DNA, whose functions were also discussed in Chapter 1, has come under further scrutiny. Scientists are discovering that the molecule plays more varied and important roles in the cell than had been previously thought, adding to speculation that it may have been the first molecule of life. Recently, biologists have discovered pieces of RNA appearing to edit DNA sequences that have dropped essential codes; other RNA molecules, called RNAi for "interference RNA" and miRNA for "microRNA," have been found to play a role in regulating gene expression. Early documentation suggests that RNAi operates within the framework of the epigenetic code described in a previous section, but it is too soon to be certain. Scientists have announced their access to an "RNA library" of chemicals that can deactivate each of a roundworm's genes, lending a new tool to computational biology, a fairly recent branch of science that emerged to compare the genome of one organism to that of another to reveal information about the role of each gene, with the ultimate goal of using similar procedures to "interview" the human genome. By comparing the nearly identical genetic circuitry of two different species like humans and mice, scientists hope to elicit the differences that make one creature become a man and the other a mouse.

There is one other place RNA is found that was totally unexpected, and that is in sperm cells. Until recently, experts thought male gametes were quite streamlined by the time they were mature, ridding themselves of most of the cytoplasm in which RNA normally resides and carrying little but the father's chromosomes. But the recent discovery of several thousand RNA molecules in each sperm raises significant questions about their role, especially in regard to fertilization and development. Some scientists are suggesting that sperm RNA is the heretofore unknown agent that unites male and female chromosomes at fertilization and jumpstarts the meiotic division of the ovum. It may also have a role in early embryonic development, at least in terms of the embryo's body plan, and its absence may explain why cloned embryos fare so poorly in comparison to those conceived sexually. The investigators also propose that sperm RNA may explain why **parthenogenesis**, a form of reproduction in which an unfertilized egg develops on its own, cannot succeed in mammals as it does in certain nonmammalian species. The theories surrounding sperm RNA, and even its very existence, are still questioned, but those who believe it exists are convinced that ongoing **microarray analyses** will confirm their speculations about the molecule's role in fertilization and embryogenesis.

OUR EVOLVING SELVES

The Evidence for Evolution and Common Descent

While genetics and molecular biology are providing insights into the human body's mechanisms of heredity and reproduction, they are also zeroing in on human evolutionary history to evaluate its evidence with greater precision. The geological records that are fundamental to much of Darwinian theory continue to reinforce it: fossils from ancient strata of the earth have been indisputably shown to bear resemblance to their fossilized descendants in the next stratum forward. Current lineages can also be traced, when fossils found in the most recent stratum resemble life forms on earth today. What's more, the fossils also depict the reality of Darwin's concept of branching descent, that all of life on earth probably arose from a single organism.

Any doubts prior to the middle of the twentieth century were put to rest by molecular analyses, when biologists discovered just how intimately present-day humans are related to their ancestors. Examining genomes and the genetic structures comprising them has tapped a wellspring of evidence. A comparison of the human genome with the chimpanzee's revealed that the human's is about 95 percent identical to the chimp's; a comparison gene for gene revealed they are even closer, with a 99.2 percent similarity. One of the most significant findings from this type of analysis is not necessarily how many genes humans share with other species, but how the differences between species hinge on a very few genes. In the case of humans and chimps, less than a 1 percent difference in genetic material accounts for a human brain three times larger than a chimp's. In December 2003, scientists announced the identification of a few of the genes that separate humans from chimps, including some involved in shaping bone structure and connecting the brain's wiring. Further analyses have shown that once these genes entered human lineage, they evolved rapidly in response to natural selection pressures that differentiate humans from chimps.

Because almost every organism of every species has the same genetic code specifying the same amino acids, whether the organism is a one-celled bacterium or a human, it has been possible to trace lineage by **biomarkers**, genetic signposts in DNA like mutations. (Those organisms having a different genetic code belong to the subkingdom "archae," a unique group that is capable of living in very extreme environments and may have been related to the first cells on earth.) Not only does the nearly universal genetic code tie species closer together and support common ancestry, it provides the chemical clues for tracking gradual evolution from primitive organisms to modern species.

What was this common ancestor? While evidence continues to redraw the animal kingdom's tree of life in terms of exactly where intermediate species fall, its main trunk seems to lead to the lowly sponge. An ancient organism dating back over 500 million years, yet similar to today's sponges, this creature may have been the first in which multiple cells communicated in a coordinated fashion to sustain the animal. With no brain, no head, and no nervous system, the sponge has nevertheless been identified as the most likely organism at the base of animal life.

As scientists continue to examine nucleotide segments of different species, the proof for common ancestry and branching descent continues to solidify. This is particularly true in the fascinating case of the HOX family of genes that determines overall body plan—the location of the head in relation to the torso or the placement of the eyes—by signaling other genes to trigger certain cellular activity at certain times. As the blueprint for the development of the organism, HOX genes have been remarkably successful in evolutionary terms for they have remained virtually unchanged, or highly conserved, for millions of years in widely divergent animal species. That these genes display only infinitesimal differences to account for the amazing diversity of animal life further supports a common ancestry.

Modern Theoretical Approaches to Evolution and Inheritance

The term *neo-Darwinism* was coined after Mendelian principles of inheritance were woven into Darwin's original theory of descent with modification to encompass the belief that natural selection acting upon genetic mutation produces evolutionary change. Since Darwin neither used the word "evolution" to describe his theory of branching descent nor lived long enough to combine Mendelian theory with his own, neo-Darwinism arose out of the modern synthesis of evolution to unite the work of the two men. Because the word could be interpreted to imply that orthodox Darwinism was somehow fundamentally changed and thus required a new name ("neo"), it is important to emphasize that the term was created to update rather than to alter original Darwinian theory.

The updating goes on as scientists continue to ponder, and sometimes solve, many of the mysteries that obscure man's evolutionary development. There has long been, for example, great controversy surrounding gaps in the fossil record that no reasoning or hard evidence has been able to explain. Some scientists claim the gaps exist simply because the "missing link" fossils haven't been unearthed yet; others, of whom Stephen Jay Gould was a major proponent, espouse the theory of punctuated equilibrium, the idea that evolution proceeds by fits and starts and is characterized by relatively rapid evolutionary change ("relatively rapid" means, in evolutionary terms, 5,000 to 50,000 years) followed by long periods of relative stability. Although Gould died in 2002, he has left a devoted following and a lively body of literature to support his theory.

Another intriguing line of thought is the notion of the "selfish gene," a concept proposed by Richard Dawkins in his book of the same name. He suggests that natural selection, rather than operating at the level of an individual or a population or even a species, instead acts at the level of the gene. Genes use the body as a "survival machine," and it is in seizing survival advantages for themselves at the expense of other genes that they are considered selfish.

Many biologists believe that parasites play prominent roles in genetic variation. People, like all living things, are host to numerous tiny organisms that are engaged in a continuing struggle with the human host for dominance. These include viruses, bacteria, even mites. Over time, some of these parasites mutate and evolve to weaken the defenses of the host. Sexual reproduction continually refreshes the host's gene pool, forcing the parasites to devise new ways to overcome genetic defenses. Theoretically, about the time they have adapted, another infusion of genetic material changes the mixture enough that the parasites must readapt. The genes that outmaneuver parasitic assaults are obviously the ones that survive, are inherited, and ultimately change the species. Many biologists believe pieces of parasitic genes hitching a ride on the human genome make up the introns, or junk DNA (see Chapter 1).

A fairly new theory presented by Lynn Margulis and Dorion Sagan in *Acquiring Genomes* states that mutation is not the primary mechanism of evolutionary change. The authors suggest that, eons ago, primitive microbes formed symbiotic relationships that allowed them to acquire genomes from one another, giving rise to new life forms in a process known as symbiogenesis. Descendant species have since evolved that either acquired new genomes via interspecies reproduction, or underwent chromosomal rearrangement during mitosis and meiosis, in a process called *karyotypic fission* or *kinetochore reproduction*, that produced new species. The sudden appearance of these species, or saltation, figures prominently in Stephen Jay Gould's theory of punctuated equilibrium.

These are but a few of the subtexts driving spirited debate among biologists, and they have been vastly simplified for inclusion here. Nevertheless, they raise fascinating questions about human evolution, heredity, and development that will undoubtedly occupy biologists for years to come, particularly as new technology permits ever more incisive exploration of cellular processes.

SUMMARY

The secrets of life are unraveling as molecular biologists explore the intricacies of cellular and genetic machinery. Yet the revelations of recent years will be rivaled by those to come as the sophisticated tools of biotechnology dig deeper into human biology and the science of life itself.

More than cells and molecules will be under scrutiny, however. Human manipulation of life has rapidly mounting religious, cultural, and social implications that policymakers and the scientific community must scramble to deal with; Chapter 7 looks at some of these.

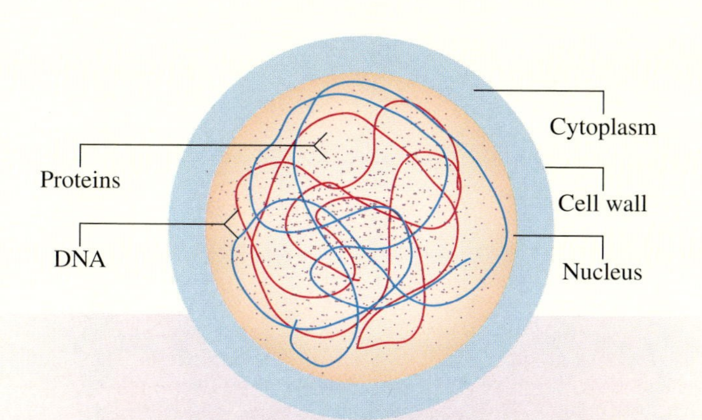

The nucleus of human cells contains chromatin, a mixture of DNA and proteins. The blue DNA is from the organism's father and the red is from the mother. This makes up the 46 chromosomes (23 pairs).

Human cell

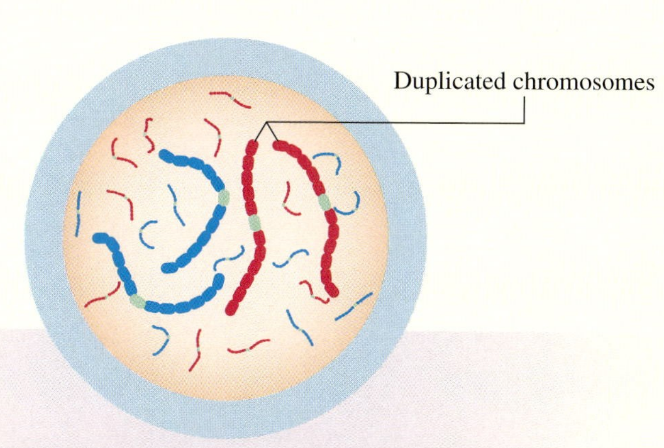

As the DNA replicates, chromatin condenses into chromosomes so that, just prior to cell division, there will be 2 copies of each chromosome, or 92 in all. (Fewer than 92 are depicted here to simplify the illustration.)

Human cell with chromosomes

Meiosis I

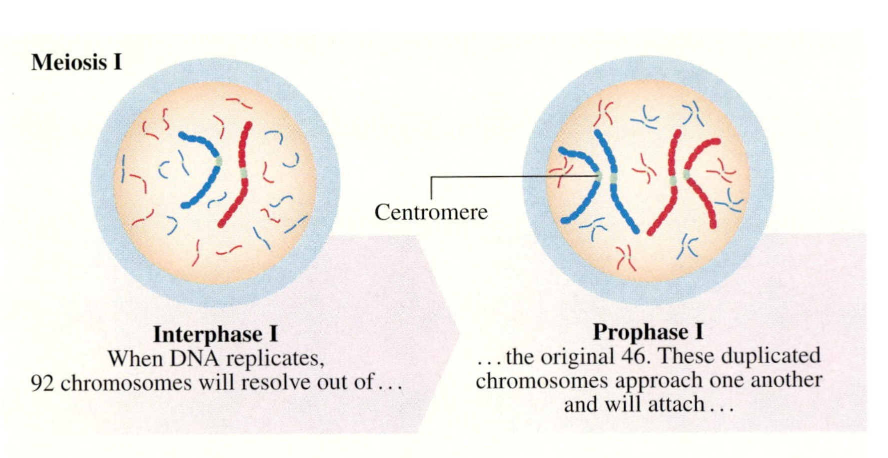

Interphase I
When DNA replicates,
92 chromosomes will resolve out of . . .

Prophase I
. . . the original 46. These duplicated chromosomes approach one another and will attach . . .

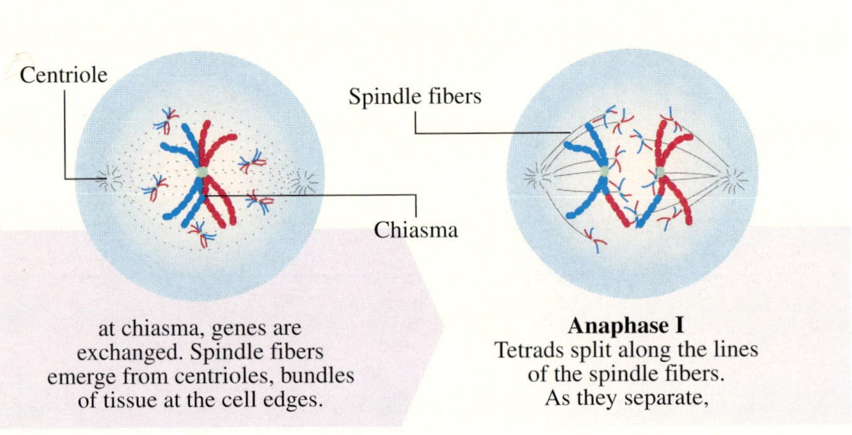

at chiasma, genes are exchanged. Spindle fibers emerge from centrioles, bundles of tissue at the cell edges.

Anaphase I
Tetrads split along the lines of the spindle fibers.
As they separate,

Interkinesis: A brief resting period between Meiosis I and Meiosis II. The nuclear

Meiosis II

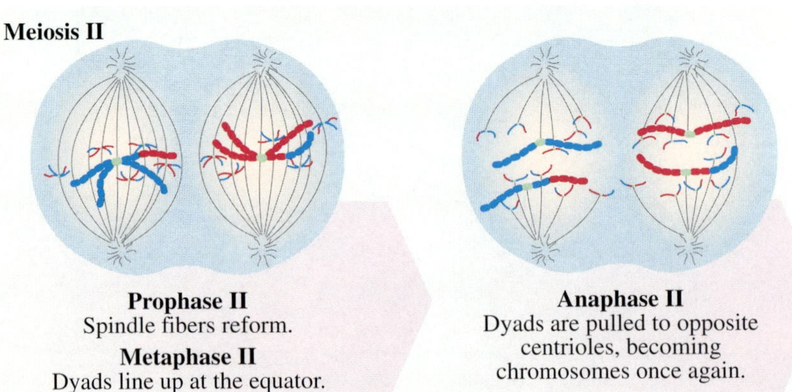

Prophase II
Spindle fibers reform.

Metaphase II
Dyads line up at the equator.

Anaphase II
Dyads are pulled to opposite centrioles, becoming chromosomes once again.

Meiosis—sperm

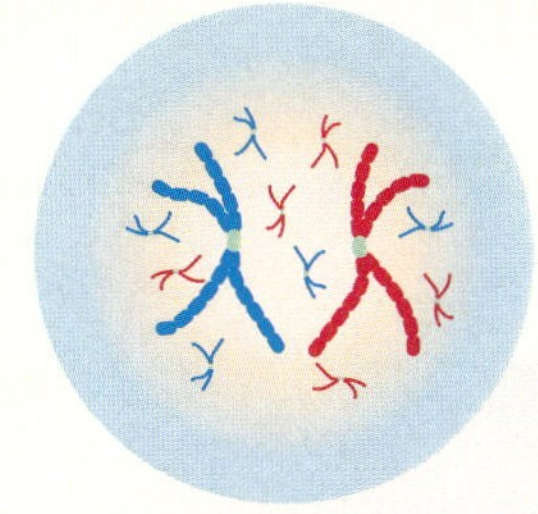

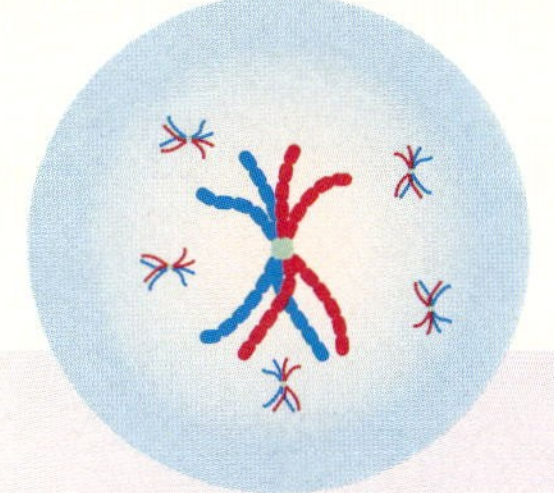

. . . at the centromeres, forming sister chromatids.

Metaphase I
Homologous sister chromatids . . .

. . . join in synapsis to become tetrads. Where chromatids overlap,

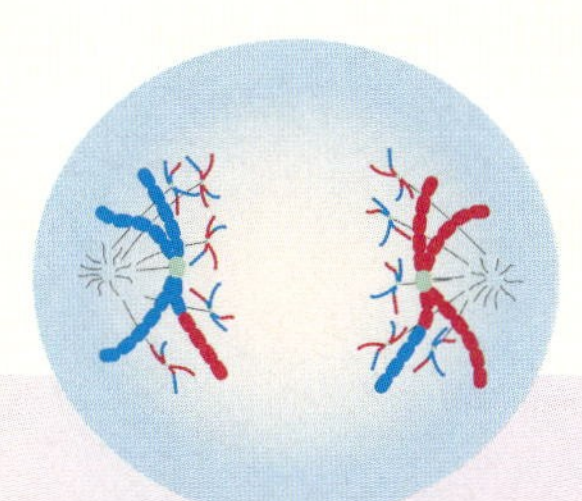

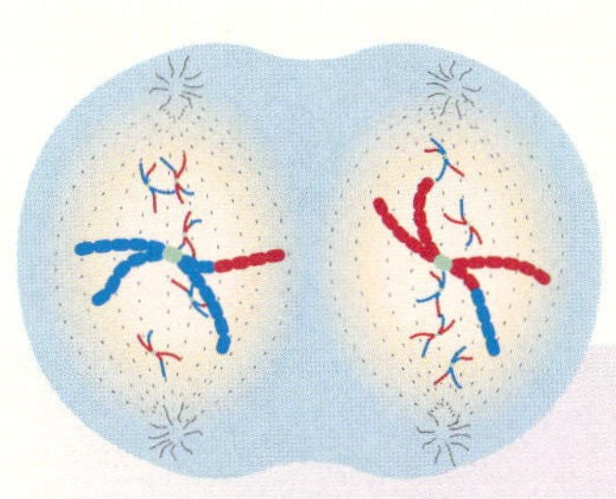

homologous chromatids are pulled to opposite sides of the cell, becoming dyads. Spindle fibers disappear.

Telophase I
Centrioles replicate, nuclear envelopes form around the dyads, and cytoplasm splits.

envelope begins to break down as chromosomes condense.

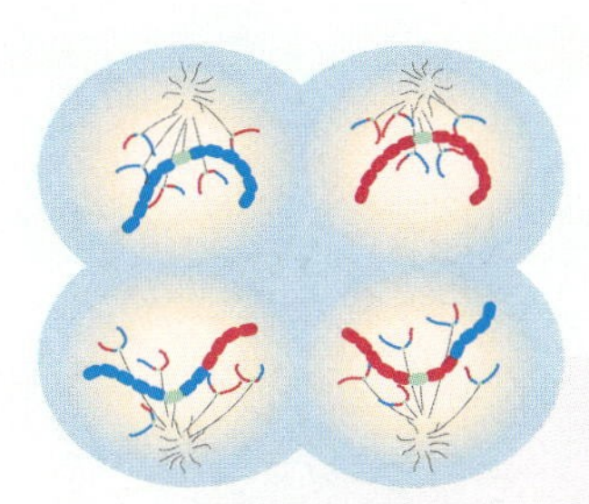

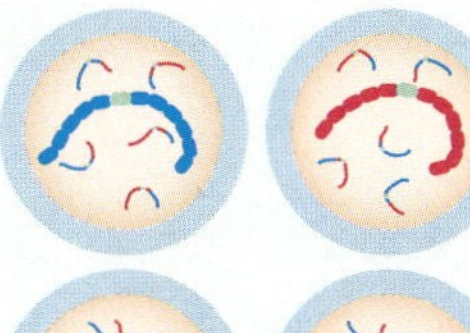

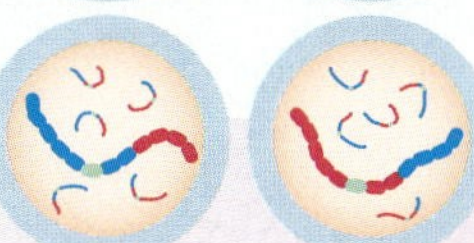

Telophase II
Nuclear envelopes form around the chromosomes,

and cytoplasmic division yields four genetically unique haploid sperm cells.

Meiosis—sperm

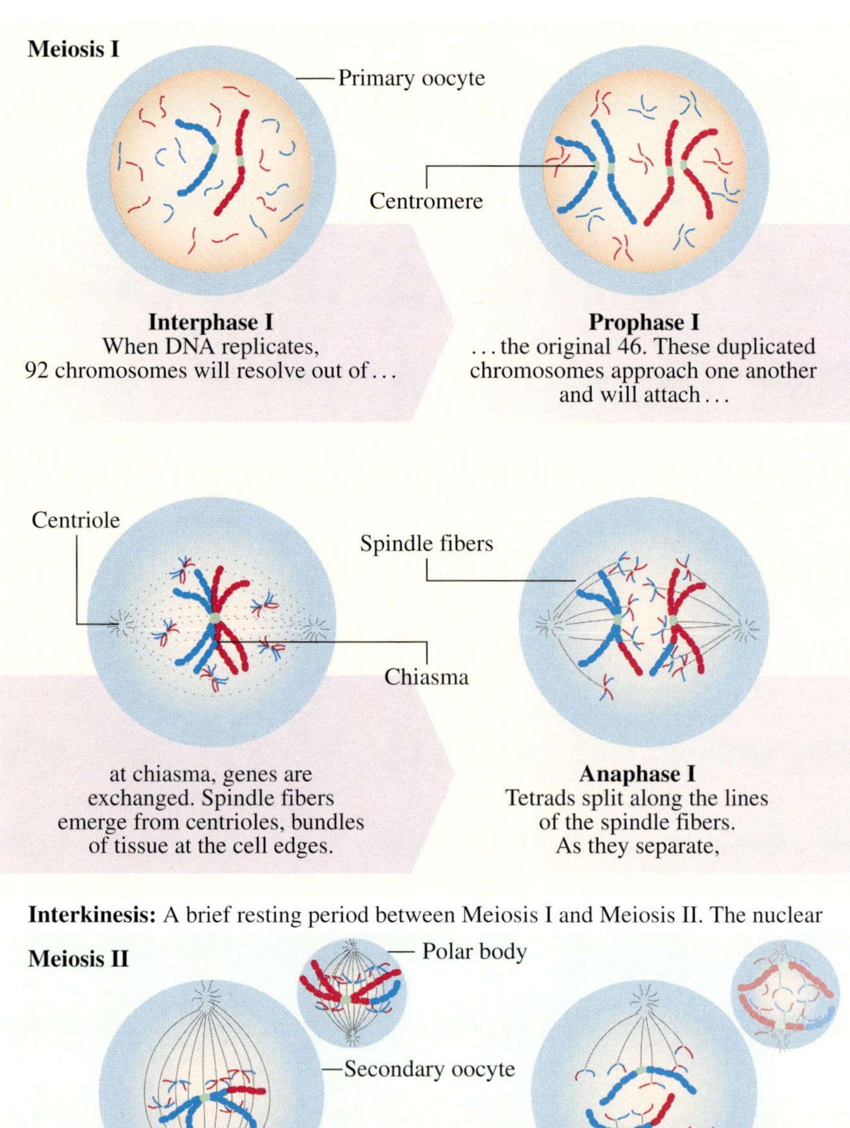

Interkinesis: A brief resting period between Meiosis I and Meiosis II. The nuclear

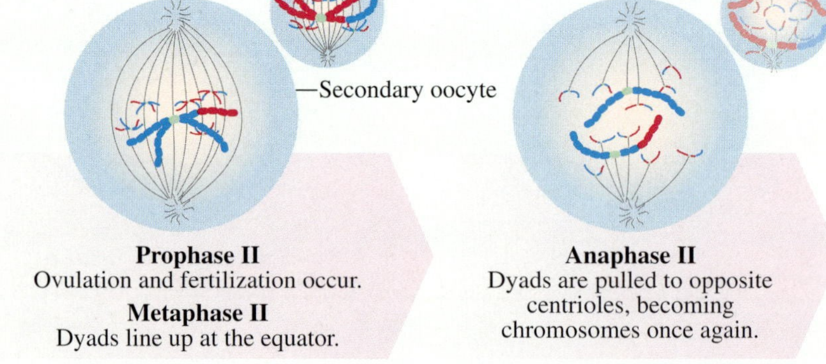

Meiosis—ovum

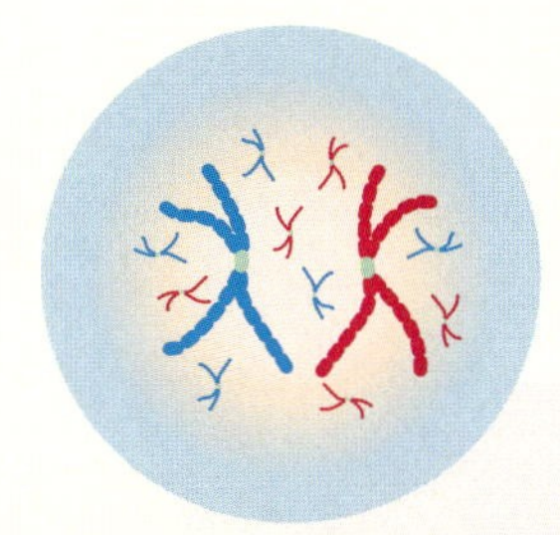

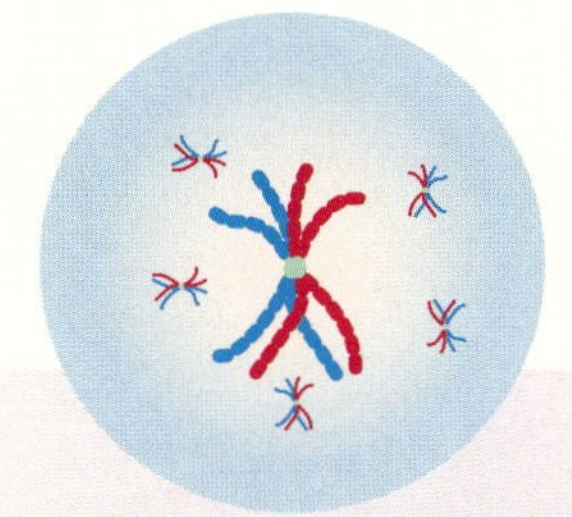

. . . at the centromeres, forming sister chromatids.

Metaphase I
Homologous sister chromatids . . .

. . . join in synapsis to become tetrads. Where chromatids overlap,

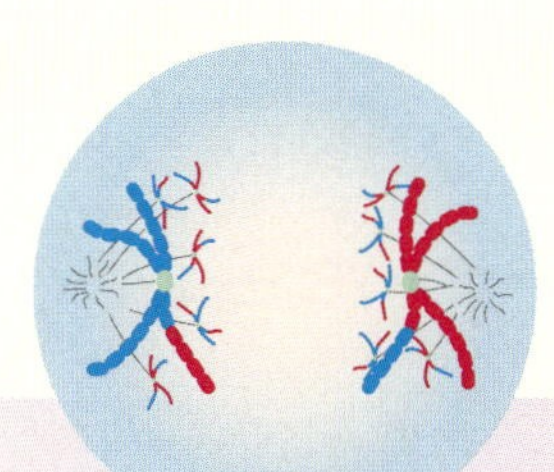

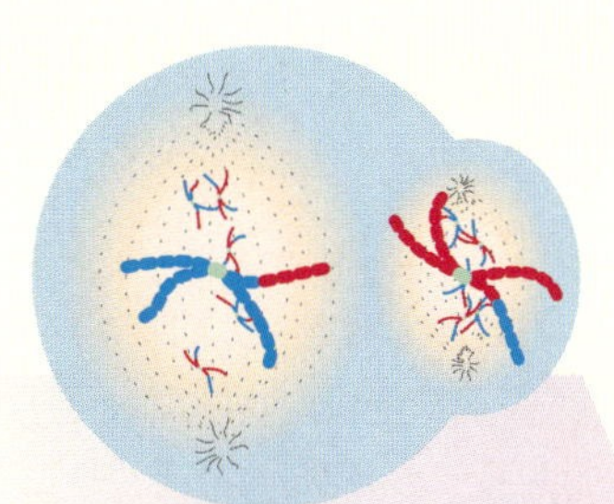

homologous chromatids are pulled to opposite sides of the cell, becoming dyads. Spindle fibers disappear.

Telophase I
Centrioles replicate, nuclear envelopes form around dyads, and the ovum begins to expel a polar body.

envelope begins to break down as chromosomes condense.

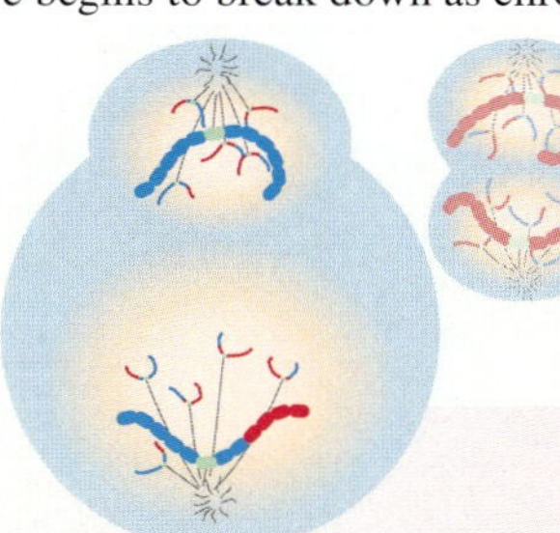

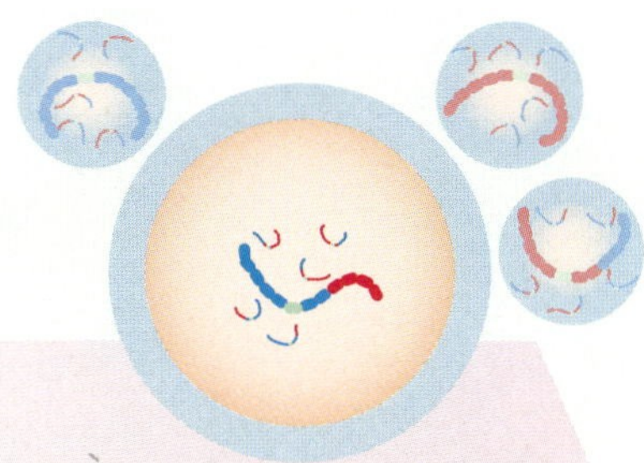

Telophase II
Nuclear envelopes form as cell division produces more polar bodies,

leaving a genetically unique fertilized ovum that will fuse with the sperm cell. The polar bodies disintegrate.

Meiosis—ovum

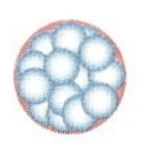

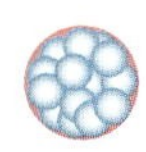

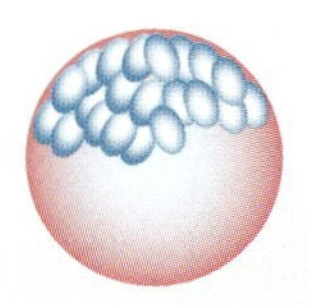

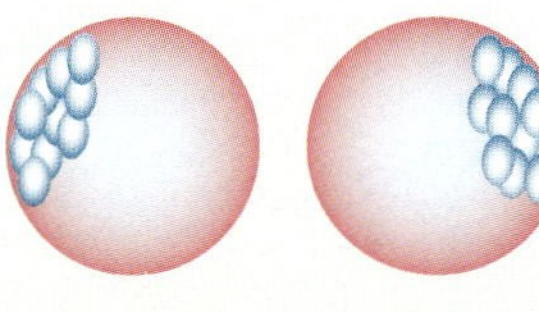

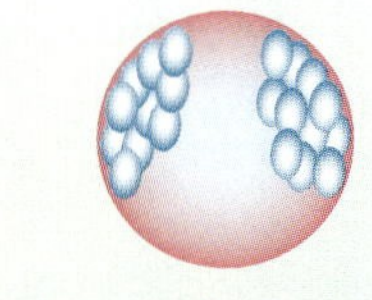

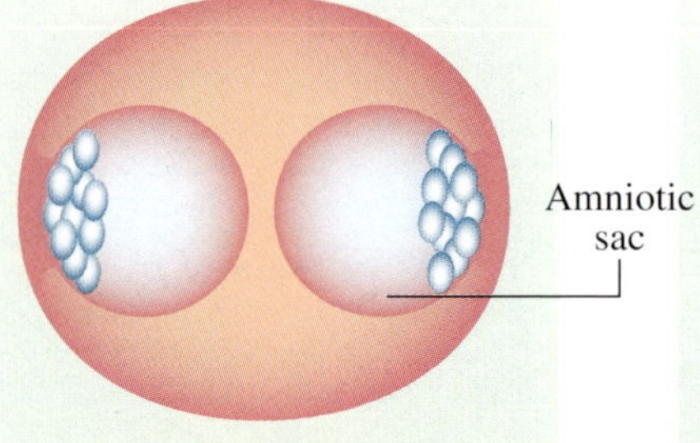

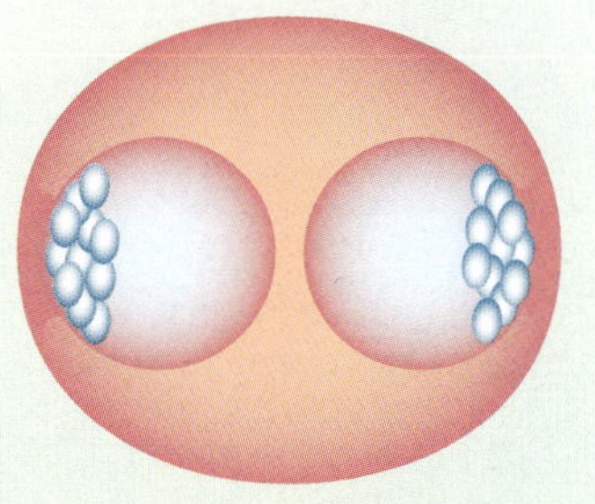

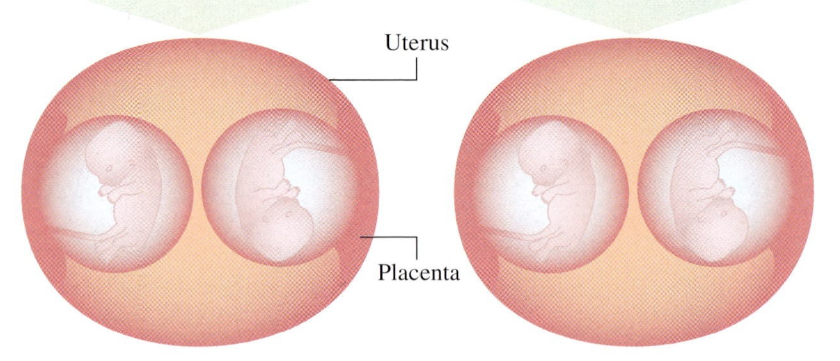

Identical twin development

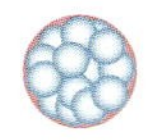

At cleavage, cells
become a single blastocyst,

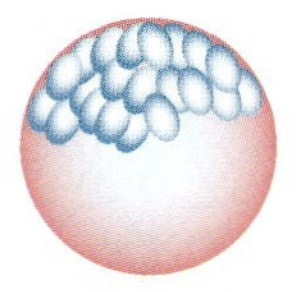

but the inner cell mass splits in two,

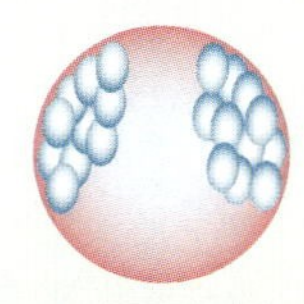

and two embryos develop,

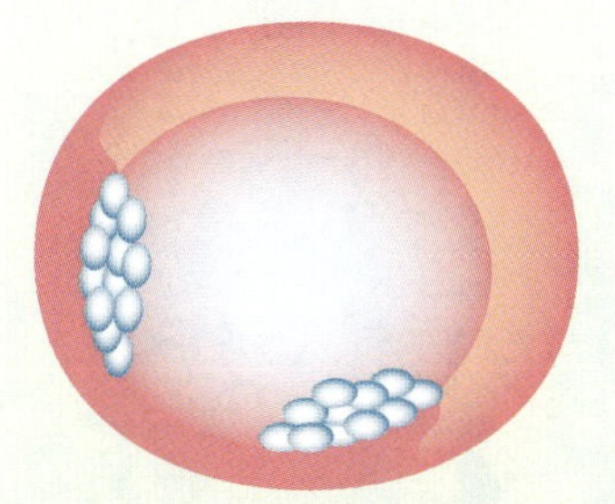

both in one amniotic sac.

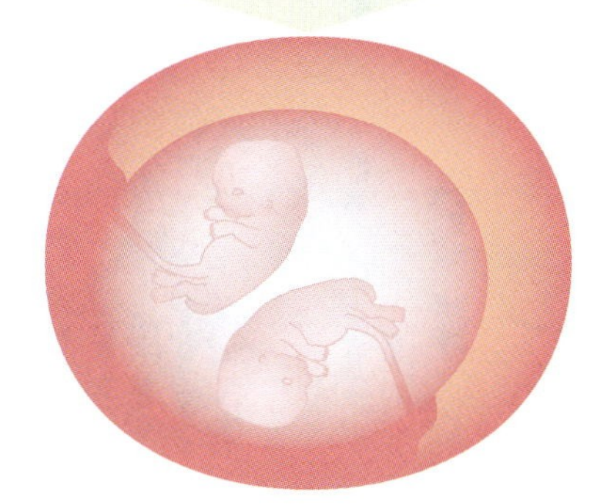

At cleavage, cells
become a single blastocyst.

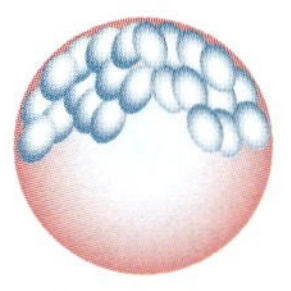

The inner cell mass starts to divide

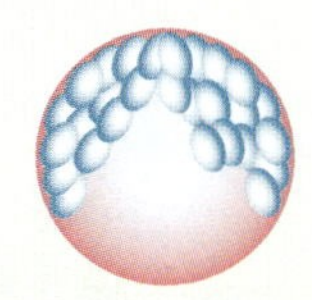

but does not completely separate,

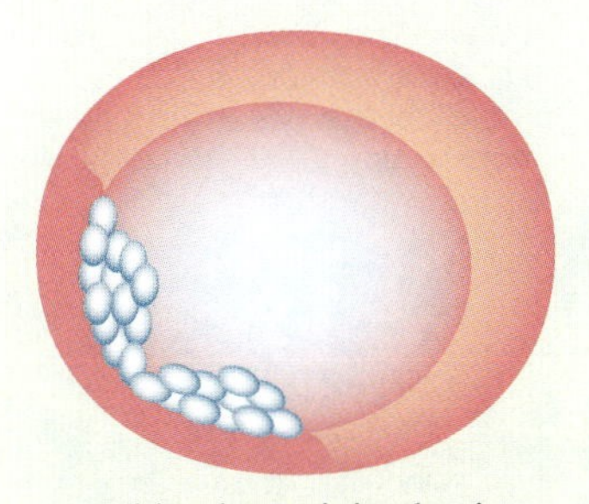

resulting in conjoined twins.

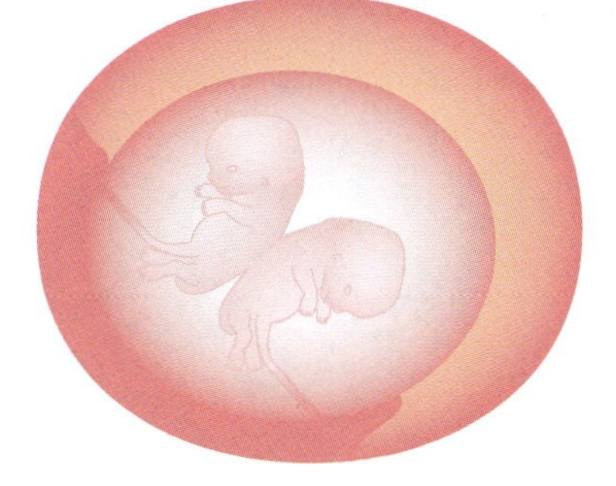

When both embryos share the same amniotic sac and placenta,
they have a dangerously high risk of tangling.

Identical twin development

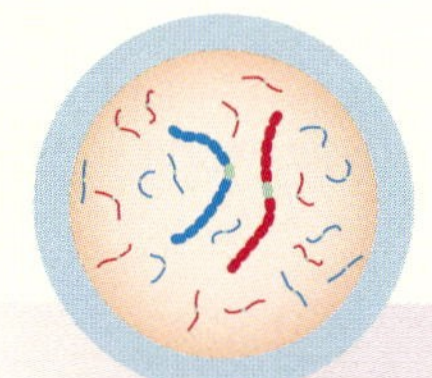

Interphase
When DNA replicates,
92 chromosomes will resolve out of . . .

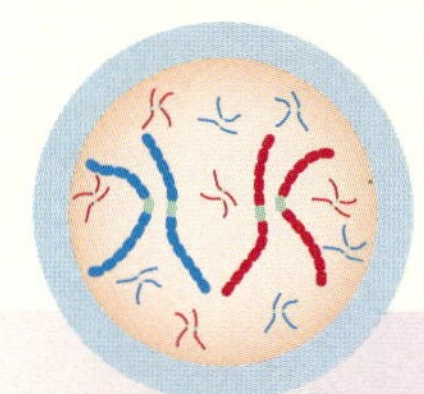

Prophase
. . . the original 46. These duplicated chromosomes approach one another and will attach at the centromeres.

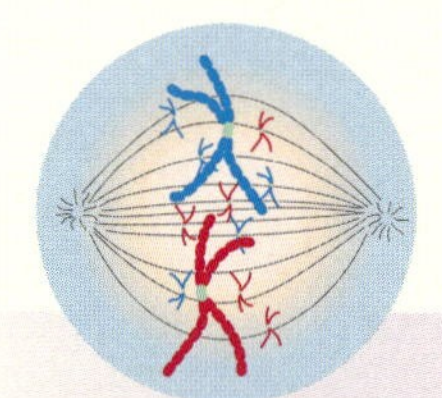

Metaphase
Nuclear envelopes break down as spindle fibers begin to form. Centromeres align on the cell's equator.

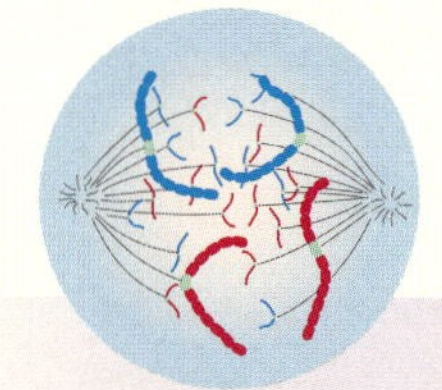

Anaphase
Pulled apart by spindle fibers, chromosomes begin to move toward opposite poles.

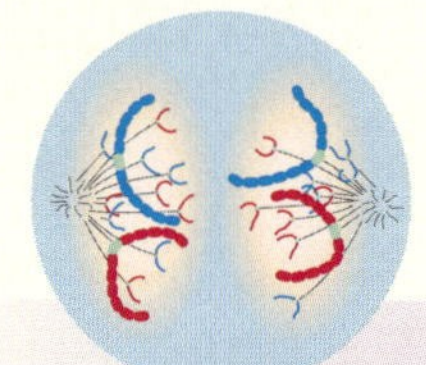

Telophase
As the separated chromosomes reach the poles, nuclear envelopes reform,

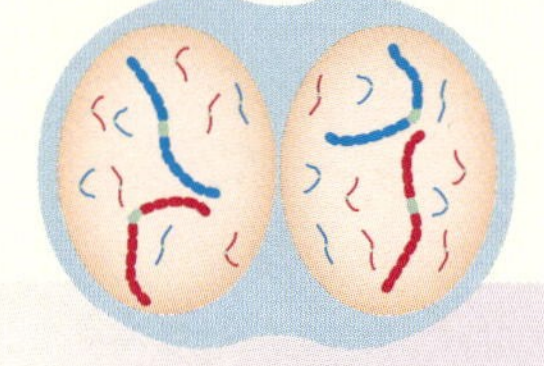

and the cell begins to divide.

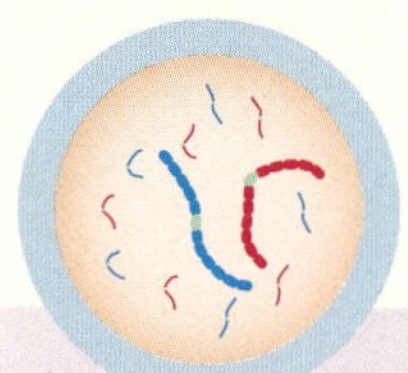
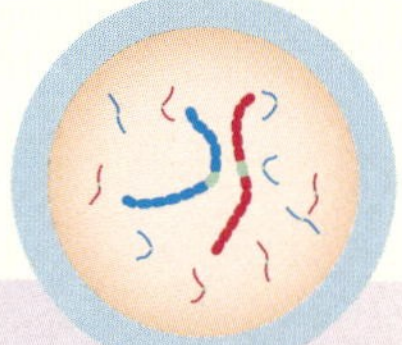

Cytoplasmic division yields two daughter cells, each having the original complement of 46 chromosomes.

Mitosis

7

New Challenges: Reproductive Therapeutics and Germline Engineering

The accelerating pace of research into genomics and proteomics is fueled by emerging **nanotechnologies** that allow scientists to manipulate the atoms and molecules of life for new purposes: treating disease, correcting genetic defects, repairing injuries, even replacing organs. Valuable experience gained in the IVF clinics that proliferated worldwide after the birth of Louise Brown in 1978 (see Chapter 6) helped not only to foster genetics research today, but also to reduce some of society's objections toward human intervention in reproductive processes. Research into stem cell and cloning technology continues to be a contentious issue, however, and the emerging potential for germline engineering—manipulating DNA in the sex cells to create heritable changes in the species—promises to be even more divisive.

Current biotechnologies that focus primarily on treating disease are discussed in this chapter, along with a brief overview of some compelling questions raised by germline engineering.

CURRENT TECHNOLOGY

Cloning

Also called *nuclear transfer*, *nuclear transplantation*, or *transnuclear fusion*, cloning means reproducing the genetic material in a cell's nucleus to create another organism, whether to retrieve its early primitive cells or to allow it to fully develop and be born a duplicate of its parent. In the cloning procedure, the nucleus from a somatic cell, perhaps a skin cell, is removed from the organism to be cloned. In the case of humans, the nucleus has forty-six chromosomes. The nucleus of an unfertilized egg, which in humans has only twenty-three chromosomes, is removed from the egg cell and replaced by the forty-six-chromosome nucleus taken from the skin cell. The egg cell, now with a new nucleus containing all the necessary chromosomes, is electrically or chemically stimulated to divide, just as a zygote conceived by two haploid gametes would. Although the cloned zygote will have the same nuclear DNA as its "parent," environmental influences and the egg cell's cytoplasmic mitochondria will impart some unique differences to the clone.

Debates surround two kinds of cloning:

- Reproductive cloning is performed to create a fully developed individual from the genetic material of another. Although one or two research groups have announced the successful birth of a cloned human, there is no evidence to verify their claims; however, many prominent biologists believe it has certainly been attempted and will, in fact, be accomplished very soon.
- Therapeutic cloning is performed to create a blastocyst from which embryonic stem cells are harvested for therapeutic purposes, such as the creation of replacement tissues or organs. In February of 2004, when South Korean scientists succeeded in cloning a human embryo and extracting a pluripotent embryonic stem cell line before the blastocyst died, the promise of therapeutic cloning became more than theoretical. Nevertheless, significant technical and ethical hurdles must be overcome before it is a practical reality.

In reproductive cloning, the zygote would gestate in the womb; in therapeutic cloning, it would grow in a laboratory until its stem cells could be extracted.

Stem Cells

EMBRYONIC STEM CELLS

As Chapter 1 explained, embryonic stem cells in a developing embryo have not yet differentiated. They have extraordinary value because they are **pluripotent cells**, cells that have the ability either to remain stem cells or to differentiate into any kind of cell in the body (also see **totipotent cells**). Moreover, they are **immortal**, and can be coaxed into reproducing indefinitely in the laboratory so that scientists have an inexhaustible supply. The source of these cells has traditionally been leftover embryos from IVF clinics.

The blastocyst does not survive removal of its stem cells, so those who regard the blastocyst as human believe harvesting the cells is like murder. On the other hand, since cloned cells that are stimulated to grow are parthenogenic in origin and do not usually survive in humans past the blastocyst stage, many feel their "natural" death renders them ethically acceptable subjects for research.

Although an inexhaustible supply of stem cells can be coaxed to grow in the laboratory, more than one embryo is needed to supply them because the stem cells derived from a single embryo have the same genetic structure. Currently, research on disease must be conducted against a variety of genomes to establish how drugs and gene therapy will affect them. In addition, cells or tissue transplanted into a patient must be compatible or the patient's immune system may reject them. The most desirable way to develop new tissues from stem cells is to clone an embryo from one of the cells of the patient needing a transplant, then retrieve the stem cells from the blastocyst that results.

Some of this technology, particularly human reproductive cloning, is theoretical and may not turn out as scientists expect. In addition, genetic imprinting discussed in the last chapter must be considered. It is possible that the cloning procedure itself prevents proper imprinting to take place and results in immediate developmental problems. This is a complex area that cannot be fully understood until a great deal more research is carried out.

ADULT STEM CELLS

These are known as **multipotent cells** and may be present in all organs of the body. As far as researchers currently know, they can differentiate only into a limited number of tissues and thus have limited therapeutic value. For example, bone marrow cells can give rise to blood cells but not to certain other kinds of cells. A further drawback to using adult stem cells is recent evidence that they tend to fuse with other tissue cells, rather than create new ones; this has very dangerous implications in terms of triggering cancerous growths or other diseases.

According to a recent report, alternatives to stem cells obtained from embryos might be those harvested from baby teeth. Scientists announced that such cells (dubbed SHED, for "stem cells from human exfoliated deciduous teeth") are capable of triggering the growth of brain and fat cells in mice, an experiment that could not be repeated with cells from adult teeth. Stem cells can also be retrieved from umbilical cords, although so far they appear to have limited potential. Due to the religious and legal controversy over the ethics of using embryonic stem cells, there is rising hope that plentiful alternate sources of these cells can be found, because many scientists insist adult cells do not offer the same promise. For the present, because the availability of federal funds to support research using the controversial cells remains questionable, many states have moved independently to support the universities, hospitals, and charitable foundations that are pursuing it.

There is more to the ethical debate than destroying blastocysts, however.

Many who oppose cloning, even for therapeutic purposes, do so on the basis of their concern for the fetus. They fear that harvesting cells from an embryo will lead far too easily to the next step—allowing an embryo to grow to the fetal stage to harvest its organs for transplantation. Scientists around the world condemn this kind of cloning in humans and most countries have voted to outlaw it. Therapeutic cloning is generally supported, however. The National Institutes of Health has stated, "the potential medical benefits of human pluripotent stem cell technology are compelling and worthy of pursuit in accordance with appropriate ethical standards." Nevertheless, the U.S. House of Representatives has passed a bill criminalizing both reproductive and therapeutic cloning, but the U.S. Senate is still deciding whether to permit the latter. Most other countries, including the members of the European Union, have approved therapeutic cloning but have overwhelmingly rejected reproductive cloning.

Because scientists are limited by how they obtain cell lines and how far they can proceed with the cloning process, much of what researchers have learned comes from their work with animals. Their results have been mixed. Dolly, a sheep born in 1997 that was the first mammal successfully cloned from an adult sheep (although it took nearly 300 attempts), appeared perfectly normal at birth. Over time, however, she developed arthritis and aged prematurely, dying at age 7—at little more than half her normal life span. Scientists still do not understand the reason for her decline, but they suspect it had to do with the lack of either genetic diversity or sperm RNA (see Chapter 6). Intriguingly, offspring conceived normally by a cloned animal do not seem to have the same health problems as their cloned parent.

THERAPEUTIC IMPLICATIONS

There are many potential applications for stem cells. They can be stimulated to grow new organ tissue to replace damaged or diseased tissues. Researchers can subject them to various drugs and other agents to determine the genetic consequences. They can be modified into packages of "cellular therapies" that can be implanted into patients, such as giving diabetic patients insulin-secreting cells or paralysis victims neural cells that might regenerate into new spinal cord nerves. One area of research rests on a recent discovery in mice that shows embryonic stem cells can develop into eggs, thus advancing research on infertility, the mechanics of ovulation, and cloning. Egg cells have even been developed from male mouse stem cells, but experts do not yet know if the resulting embryos will develop normally.

If scientists can test drugs, treatments, and techniques on a wide variety of genetically different stem cells, they will, in theory, produce effective medications, tissues, and gene therapies that will work for a broad spectrum of genetic profiles. Ultimately, when they can map individual genomes easily and inexpensively by sophisticated scanning equipment, scientists expect to be able to gear treatments to each person's genetic makeup without the monumental problems that tissue and organ rejection pose today.

Genetic Testing

Completing the "finished" sequencing of the human genome (accurate to one error in 10,000 letters) in April 2003 has given scientists a 99.99 percent complete map of the human genome, making genetic testing a dramatically more useful tool. DNA testing or gene profiling to map an individual's genome can easily detect single-gene diseases like Huntington's chorea or sickle-cell anemia (see Chapter 1), but when disease is caused by a complex of interacting genes, it is much more difficult. Nevertheless, rapid progress is being made in technologies that can identify the patterns of gene activity leading to disease, and in procedures by which assessments of disease potential in populations can be made. Many scientists anticipate that when a person is born, his or her genome can be analyzed so that lifetime preventive measures are instituted to ward off diseases to which the infant appears to be genetically susceptible.

Although genetic testing is not 100 percent accurate and is fairly expensive, four different procedures are becoming routine:

- Prenatal diagnosis tests the fetal genome for diseases like Down syndrome.
- Newborn screening is often performed to detect such conditions as congenital hypothyroidism, which can be readily treated once diagnosed.
- Late-onset disorders are diseases that arise in one's lifetime; if the disease is due to a combination of genetic and lifestyle factors, as heart disease often is, detecting genetic predisposition can be lifesaving.
- Preimplantation genetic diagnosis (PGD) is more and more frequently used in IVF clinics to detect abnormalities in any of the embryos fertilized in the lab for later implantation; defective embryos are destroyed.

Gene Therapy

Still experimental as of early 2004, this method of treating disease at its molecular origin generated great excitement when it was first conceived, but the reality has not been as promising as hoped. The technique of introducing therapeutic strands of DNA into somatic cells to replace faulty DNA that carries defective genes has, until recently, depended on viral vectors, pieces of virus that transport the therapeutic genes into the body where they are capable of entering the cell. Once the new section of DNA has been integrated, every cell arising from that cell will carry the therapeutic genes that should ultimately overcome the effect of the faulty genes. But sometimes the DNA inserts itself in the wrong place, a serious problem that has led to the development of cancer in patients whose therapeutic DNA inadvertently activated nearby cancer-promoting genes. Newer techniques for inserting

DNA are making it easier to pinpoint the exact location where it will be positioned, but the long-term results of gene therapy are still questionable. There is too much to be learned about how genes control protein manufacture for experts to predict with certainty what the side effects will be, but the necessary knowledge is accumulating at an exponential rate.

GERMLINE THERAPY AND ENGINEERING

Gene therapy should be distinguished from germline therapy; in the latter, therapeutic genes are inserted into the germ cells (sperm or eggs), and thus would be inherited by offspring. Even if the technique is perfected and its potentially harmful side effects eliminated, germline therapy will probably not become a reality if germline engineering as a whole is rejected. There are grave concerns that once society permits tampering with the genome for therapeutic purposes at the level of the germ cell, the pressure to permit similar procedures for other than humanitarian purposes will be insurmountable. In particular, bioethicists suggest, parents may insist on bioengineered children even if the reasons have little to do with health.

The concept of "designer babies" is a good example. In a society that allows defective embryos to be destroyed in IVF clinics and therapeutic DNA to be inserted into somatic cells, many argue that germline therapies that can erase genetic problems or deliver life-enhancing qualities to their children should certainly be permissible. But some parents might go further, wishing to choose their baby's gender. Others might want their child to have a cheerful disposition or excel academically. Still others envision their child as a star on the athletic field. The dilemmas such issues might raise for physicians, bioethicists, legal scholars, and many others could be formidable. Some of them include:

- determining if parents should be allowed to "design" their children;
- evaluating the philosophical cost of inserting nonhuman genes into human DNA;
- assigning responsibility for experiments in germline engineering that go wrong;
- identifying the legal status of anyone whose DNA is no longer human;
- protecting privacy when individual genomes are widely accessible;
- structuring an economy that can sustain people who live a very long time; and
- dealing with cultural issues raised by human-to-cyborg interactions.

It is obvious that germline engineering has ramifications that extend well beyond the scope of reproductive biology and thus beyond the scope of this volume. But it is no coincidence that along with **reprogenetics**, terms like **posthumanism** and **artificial intelligence** increasingly appear in scientific literature. Responding to the needs of those who must grapple with myriad issues arising from genetic research, the Human Genome Project established an Ethical, Legal, and Social Implications (ELSI) Program in 1990 to begin to address them. The world's largest such organization, it is administered by both the National Institutes of Health and the Department of Energy, and is reevaluated regularly to ensure that its guidance is relevant to sensitive and ever-shifting concerns. Unfortunately, however, the legal implications of many aspects of genetics technology are not always clear. There are already bewildering patent laws surrounding genes and gene sequences that will be tested, revised, and complicated as different companies seek to protect their research investments.

SUMMARY

Stem cell research, gene and germline therapies, and cloning, and the new technologies that accompany them, represent a new frontier in biology that is changing the way scientists study the human body and the way physicians treat its injuries and diseases. The breadth of knowledge gained in the last fifty years is especially apparent when compared to a timeline of breakthroughs in reproductive medicine since ancient times. The next chapter offers such a retrospective, and sets the stage for discussions about the reproductive disorders medical science confronts today.

8

Anatomy and Surgery: A Retrospective

During prehistoric and ancient times when abdominal surgery usually meant death, most people had to do without. A pregnant woman unable to give birth because her fetus lay obliquely in the womb was often sacrificed to rescue the child, although her life might be saved if the fetus was dismembered and removed through the vagina. Ecclesiastical writer Tertullian recorded between 200 and 250 CE that "the limbs are dissected within the womb," whereupon the entire fetus is "violently extracted." Cave drawings depicting mothers in labor do not reveal whether obstructed fetuses of prehistory met a similar fate, but there is every reason to believe their mothers suffered unimaginable agony.

Although modern medicine owes a great deal to the teachings of scholars from ancient civilizations, their ignorance of human anatomy prevented medical science from progressing very far on the surgical front. In the mid-1500s, when Andreas Vesalius (1514–1564) published a comprehensive anatomy text based on dissections of human rather than animal cadavers, the situation began to change. As reliable anesthetics were introduced in the 1800s, abdominal and reproductive surgeries rapidly increased.

Following this brief narrative is a retrospective chronology illustrating how knowledge of human anatomy led to surgical techniques that, in turn, triggered greater insights into anatomy, a synergy that is magnified by today's technological revolution.

ANCIENT AND MEDIEVAL PERIODS

Chapter 5 pointed out that classical civilizations produced great thinkers whose sophisticated knowledge of medicine was astounding for their time.

No doubt drawing on ancient texts from Egyptian, Indian, Arabic, Greek, and other cultures, Soranus of Rome (98–138 CE), known as the founder of obstetrics, wrote at length about menstrual difficulties, infertility caused by genital tract inflammation, pregnancy, and advanced procedures such as positioning a fetus for delivery and packing the uterus in the event of hemorrhage.

What was also notable about the ancients was their absence of surgical expertise. Respect for the sanctity of the human body, a prevalent theme, prevented in-depth exploration of internal anatomy. This meant that surgery was primarily based on wound management and bloodletting. Although Hippocrates mentioned various nonsurgical treatments for uterine **prolapse**, no authenticated vaginal hysterectomy, in which the uterus is surgically removed, had ever been attempted to treat prolapse until 1507. Surgery on men was limited as well, although circumcisions and removal of the testicles for cultural purposes or enslavement had taken place for centuries.

The "Wound Man" is a medieval illustration of a grown male displaying the typical injuries physicians were capable of treating. The drawing served as an anatomical guide until **postmortem** examinations, such as the one Jacopo Berengario da Capri (ca. 1460–ca. 1530) reportedly performed on a deceased pregnant woman, revealed correct anatomy. Seizing the opportunity to depict the human form correctly, many Renaissance artists including Leonardo da Vinci (1452–1519) produced detailed sketches of the internal organs that are remarkably accurate even by today's standards. When Andreas Vesalius published his masterpiece *De humani corporis fabrica* (*On the Fabric of the Human Body*) in 1543, he debunked many of Galen's centuries-old pronouncements about the human body and produced the world's first realistic anatomy text. Believing it necessary to observe the body itself, Vesalius circumvented Church dictates by secretly obtaining cadavers from local cemeteries for dissection. His accurate depictions of male and female reproductive organs were early roadmaps for many medieval surgeons, including Ambroise Paré (1510–1590). Acclaimed for his two books on surgery that included discussions of obstetrics and infertility, Paré, who trained as a barber-surgeon and whose contributions were thus initially dismissed by physicians, was the first to surgically repair a vaginal partition (septum) in a woman unable to conceive.

THE RENAISSANCE TO 1900

Although Vesalius' *De fabrica* and his observation-based scientific method transformed medicine, it was a gradual process. At the time Vesalius published his work, William Harvey's groundbreaking work on blood circulation was yet to come, and Regnier de Graaf (1641–1673), who identified the

Graafian follicle (but mistook it for the ovum), had not even been born. Female midwives continued to preside over bedside childbirth, and it would be three centuries before Ignaz Semmelweiss (1818–1865) could prevent puerperal fever from stalking new mothers in hospital maternity wards.

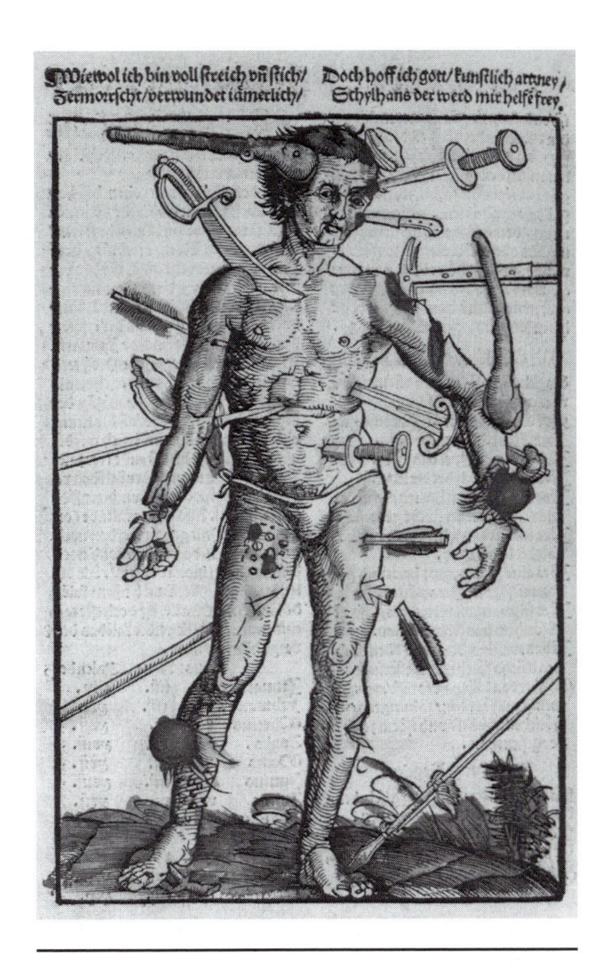

This reproduction of a medieval woodcut shows the kinds of wounds that surgeons of the day were capable of treating. By permission of the British Library. C.31 m.12 pg XVIIIv.

Nevertheless, knowledge of abdominal and reproductive anatomy accumulated throughout the 1700s. As surgical procedures to save mothers and fetuses became more advanced, "man-midwives" began to replace female midwives. One of them, William Hunter (1718–1783), was a renowned personal physician to royalty who went on to establish a prominent obstetrics and anatomy school in London. His brother John (1728–1793), also a physician to royalty, experimented with human insemination and became a master of anatomy at Surgeon's Hall in England. A contemporary of the Hunters, William Smellie (1698–1763), a Scottish obstetrician who taught man-midwives, published three texts on midwifery and introduced the proper use of forceps into the practice of obstetrics.

Ephraim McDowell (1771–1830) made history in 1799 when he performed the first **ovariotomy** on a woman suffering from a very large ovarian cyst; by 1872, the procedure was so routine it was performed on healthy ovaries in an attempt to relieve certain female neuroses that were believed, in many cases, to originate in the female reproductive organs. By the time James Marion Sims (1813–1883) perfected a surgical procedure for repairing tears in the vagina (fistulas) arising from difficult labor, hysterectomies had already become commonplace. Generally considered the father of gynecology, Sims established the Women's Hospital in New York in 1853.

Up to 1867, castration had been the only treatment available to reduce a severely enlarged prostate, but that year Theodor Billroth (1829–1894) surgically removed a man's diseased prostate through an incision between the patient's scrotum and rectum in a procedure known as radical perineal prostatectomy. It remained the surgical standard until 1949, when Dublin physician Theodore Millin introduced a less radical retropubic technique that gave the surgeon access to the prostate and lymph nodes through the abdomen. By 1900, genitourinary surgeons were also experimenting with removing the vasa deferentia (vasectomy) as an adjunct to treating enlarged prostates.

Tumors and other disorders of the breast had, of course, been plaguing women (and even some men) for centuries, and in the absence of other available treatment, badly diseased breasts had been removed. Although some reports suggest various combinations of herbs, alcohol, laudanum, morphine, or opium were used to relieve pain, there are harrowing accounts in the scientific literature of **mastectomies** performed on women with no anesthesia at all. Fortunately, this state of affairs ended in the middle of the 1800s when ether, nitrous oxide ("laughing gas"), or chloroform were used to anesthetize patients. Towards the end of the century, as chloroform proved more toxic than ether, its use gradually decreased and was discontinued in the early 1900s.

THE TWENTIETH CENTURY . . . AND BEYOND

Early in the twentieth century, the use of **endoscopic** tools that allowed physicians less invasive access to internal organs became more widespread. The first such instruments, used in the middle of the 1800s, relied on an open flame for illumination and a series of mirrors to reflect images, but, by 1911, Bertram Bernheim (1880–1958), who introduced the laparoscope, was able to use ordinary light. A small tube inserted through the abdomen that permits visualization of internal organs, the modern laparoscope is frequently used in tubal ligation, hysterectomies, and prostate surgery.

Until the mid-1900s, subtotal hysterectomies, in which only the uterus is removed, were most common, but once it became increasingly clear that the cervical stump remained vulnerable to disease, total hysterectomy, removal of the uterus and cervix, became more popular. In radical hysterectomies, surgeons remove the ovaries and surrounding tissues as well, but they try to avoid this procedure in younger women to prevent premature menopause.

Although prostate removal to treat disease has saved many lives, unfortunate side effects of the procedure left most men impotent (see Chapter 9) and about a quarter of them **incontinent**. In 1980, an American, Patrick Walsh, demonstrated that critical blood vessels and nerves lay outside the prostate rather than inside it, and thus could be preserved during the gland's removal. His discovery formed the basis of a nerve-sparing surgical technique that has avoided such consequences.

After the 1950s, genetics transformed medicine, changing how doctors would understand, diagnose, treat, and prevent many diseases. Beginning with the Nobel Prize won by Thomas Hunt Morgan in 1933 for his discovery of chromosomes' role in heredity, subsequent awards acknowledged similar contributions to the understanding of genes in health and disease. Some of these include:

- 1946, to Hermann Joseph Muller (1890–1967), for discovering how x-rays cause genetic mutations;
- 1958, to George Wells Beadle (1903–1989) and Edward Lawrie Tatum (1909–1975), for discerning how genes regulate cellular chemistry; and to Joshua Lederberg (b. 1925), for explaining genetic organization and recombination in bacteria;
- 1995, to Edward B. Lewis (b. 1918), Christiane Nüsslein-Volhard (b. 1942), and Eric F. Wieschaus (b. 1947), for discovering how genetic factors affect embryogenesis; and
- 2002, to Sydney Brenner (b. 1927), Robert H. Horvitz (b. 1947), and John E. Sulston (b. 1942), for clarifying how genes program the death of cells to help regulate organ development.

These discoveries, and many others, have led to dramatic improvements in diagnosing diseases and in approaches to both prevention and treatment. As the timeline shows, many physicians already sit at remote computer screens to control robotic "surgeons" in the operating theater. In just a few years, according to biotechnology and nanotechnology experts, it is very likely that microscopic robots will be scurrying throughout the human body to repair diseased and injured organs from within.

Date	Anatomy	Pregnancy and Fertility	Surgery
Prehistory	Rock drawings of women giving birth indicate people had some understanding of internal organs.	Skeletons are found with fetuses wedged in the pelvis, indicating that many women probably died in childbirth.	*Note: Few abdominal surgeries are performed before the 18th century and then only in extreme emergencies to save the fetus (but sacrifice the mother).*
Antiquity (10,000 BCE–1100 CE)	Respect for the body prevents first-hand observation of human anatomy; most information is gleaned from animals. The Greeks can describe the uterus, but think it wanders around in the abdomen, causing hysteria (the source of the word "hysterectomy"). Galen proposes his theory of inverse symmetry in which he claims the vagina is an inverted penis; other female organs are equated with those of males, but Galen has no explanation for the clitoris. Soranus of Rome explains the anatomical difference between the uterus and the vagina; he also documents pelvic abnormalities and recommends measures to prevent tearing of pelvic organs during childbirth.	Early manuscripts show that breech presentations (feet first) are understood and directions are given showing how to turn the child in the womb. Barrier methods of contraception are used that are made of herbs and crocodile dung; Aristotle proposes that cedar oil or frankincense oil makes a good spermicide. In Greece, Herophilus discovers the prostate gland; Soranus of Rome writes *Gynaecology*, in which he exhibits a remarkably sophisticated scientific understanding of reproductive biology. For easing childbirth, Soranus devises a birthing stool with an opening in the seat and support structures on which the laboring mother can rest her arms and back; different versions of the stool or chair arise through the centuries. Often, women with no access to birthing stools are supported in someone's lap while the midwife assists the delivery from below. Soranus advises tests for newborns that are similar to today's that evaluate babies' overall health and response to touch.	In Egypt, circumcision is mentioned in an ancient papyrus; some surgery is performed on bones; first female circumcision is recorded; castration is routinely performed for the purposes of ritual, enslavement, or to preserve the male soprano singing voice. Surgeries up until the middle of the eighteenth century usually concern lancing boils, bloodletting, or setting bones. A woman might be cut open to retrieve her fetus, but it usually costs her life; sometimes the fetus is dismembered through the vagina in order to save the mother. Soranus describes the removal of a prolapsed uterus and is reported to have performed the first vaginal hysterectomy to treat this disorder. However, the first authenticated surgery of this kind was performed by Jacopo Berengario da Capri in 1507; rarely did women survive these procedures prior to the Middle Ages.

Date	Anatomy	Pregnancy and Fertility	Surgery
1100–1300	The dissection of human corpses gives scientists a view of and comprehension of human anatomy. Distinguished schools of anatomy are established in both Italy and France.	Birth is an all-female activity, with female midwives rising in status.	
1300–1600	Jacopo Berengario da Capri performs postmortem examinations, including one on the body of an executed pregnant woman, advancing understanding of female anatomy. Renaissance painter Leonardo da Vinci beautifully depicts human anatomy, especially embryogenesis and the urinary tract. Anatomy theaters are established in which students view autopsies and receive instruction. Andreas Vesalius publishes *De humani corporis fabrica (On the Fabric of the Human Body)*, correcting Galen's errors and putting to rest the misperception that women's organs are the inverse of men's; his work becomes the basis of subsequent anatomical studies. Jacob Rueff's *De conceptu et generatione hominis* uses anatomical pictures that revise and correct some of Vesalius' illustrations; the book is translated until well into the seventeenth century as a guide for midwives, covering conception to sterility.	Jacob Rueff publishes a book on conception and birth for both students of medicine and anatomy as well as midwives. Many physicians believe infertility is God's punishment for enjoying sex. Frenchman Arnaud de Villeneuve claims that obesity helps cause infertility because "fat suffocates the seed of man." Italian anatomist Bartolomeo Eustachio may have anticipated techniques of artificial insemination when he instructs men to place a finger in a woman's vagina after intercourse in order to facilitate conception. French surgeon Ambroise Paré urges dilation of the cervix as a correction for infertility. Forceps are introduced by the Chamberlen family who uses them exclusively, keeping them secret from others for nearly four generations.	Renaissance surgeon Ambroise Paré, who wrote two books on surgery, is the first to operate on a vaginal septum (membranous partition).

Date	Anatomy	Pregnancy and Fertility	Surgery
1300–1600 *(continued)*	Bartolomeo Eustachio helps delineate the uterine vessels.		
1600–1700	Regnier de Graaf discovers the Graafian follicle and gives "ovary" its name; however, he believes the follicle to be the ovum. Based on von Leeuwenhoek's microscopic examinations, Nicholas Hartsoecker depicts spermatozoa in illustrations, but he states that each sperm contains an already-formed embryo.	De Graaf postulates the importance of the ovum in reproduction. Midwives begin attending obstetrics classes. François Mauriceau writes *Treatise on the Illnesses of Pregnant Women* in which he debunks the belief that the pelvic bones actually separate during childbirth. *De Sterilitate* is published by Martin Naboth in which he discusses how ovarian scarring and tubal blockages might be responsible for infertility.	Surgeons attend childbirth only when it is necessary to extract a dead fetus. Hendrik van Roonhuyze publishes a book describing surgical procedures concerning ectopic pregnancy and Caesarean sections.
1700–1800	William Hunter, obstetrician and physician to the British queen, becomes a Professor of Anatomy at the Royal Academy; publishes *Anatomy of the Human Gravid Uterus* that is a groundbreaking depiction of pregnancy and fetal development.	Physicians identify various anatomical abnormalities that are responsible for infertility. John Hunter's early attempts at human insemination result in the birth of a child. Childbirth begins to be regarded as a natural event, rather than one of illness; women are encouraged to deliver in fresh-air filled rooms and breastfeed their babies.	Male physicians begin to supplant midwives, especially since they carry surgical tools needed for troublesome deliveries. Scotsman William Smellie sets up surgical-obstetrics schools and helps train surgeons. Mortality rate for hysterectomies is about 90 percent.
1800–1850	The first endoscopic instrument that visualizes the genitourinary tract is developed.	Precursors of the diaphragm and condom become available for birth control.	Mastectomies are performed without anesthesia. The first successful hysterectomy for cervical cancer is performed, and hysterectomies become more common.

Date	Anatomy	Pregnancy and Fertility	Surgery
1800–1850 *(continued)*			A major breakthrough in gynecologic/obstetric surgery occurs when Ephraim McDowell successfully removes a woman's ovarian cyst. James Marion Sims begins to repair vaginal fistulas (tears that commonly develop in the vagina or bladder after difficult deliveries) and helps establish gynecology as a specialty. Hysterectomies become safer and are increasingly performed to treat a variety of disorders. Anesthesia is routinely used in surgeries by 1840 and sterile procedures are observed in hospitals. Sir James Simpson introduces chloroform and ether for childbirth, hysterectomies, and other surgery. The first successful subtotal hysterectomy (only the uterus is removed) is performed.
1850–1900	The existence of sex hormones is discovered by Viennese gynecologist Emil Knauer.	James Marion Sims carries out numerous artificial inseminations. The discovery that conception occurs when the sperm enters the egg is made, confirming that the fusion of two gametes results in fertilization. More birth control devices become available. The modern version of the diaphragm is invented.	James Marion Sims founds women's hospitals in America and introduces new gynecological surgical techniques. Several physicians develop surgical procedures to treat the ovaries, Fallopian tubes, and the uterus. The first hysterectomy on a pregnant woman is performed.

Date	Anatomy	Pregnancy and Fertility	Surgery
1850–1900 *(continued)*		The first case of donor insemination is announced.	The first radical perineal prostatectomy is performed to treat benign prostatic hypertrophy (BPH). The first radical hysterectomy for cervical cancer is performed. Robert Lawton Tait performs the first successful surgery for ectopic pregnancy; his groundbreaking surgical exploration of the abdomen made him the father of modern abdominal surgery. The first radical hysterectomy for uterine cancer is performed.
1900–1950	The Pap smear to detect cervical cancer is introduced.	The first urine test for pregnancy is developed. The world's first birth control clinic opens. Margaret Sanger advises women on "birth control," a term she coins herself. Scientists devise the "rhythm" method of birth control. Endocrinologists postulate that hormone therapy may be useful in birth control. The Rh factor of blood is discovered, allowing obstetricians to avoid life-threatening Rh-associated problems for newborns.	Genitourinary surgeons begin performing vasectomies to relieve severe prostate enlargement; in extreme cases, castration is performed to treat the disorder. Caesarean sections become safer. The laparoscope is introduced for surgical procedures (hysterectomy). Minimally invasive endoscopic surgery replaces laparoscopic surgery in some cases. Hugh Young performs the first perineal prostatectomy specifically for the treatment of cancer. Laparoscopic surgery is introduced to the United States.

Date	Anatomy	Pregnancy and Fertility	Surgery
1900–1950 *(continued)*			Hysterectomy mortality rate is reduced to about 18 percent. Injectable forms of anesthesia become available. A retropubic approach to prostate surgery is used that allows removal of the lymph nodes.
1950–2000	The discovery of the pituitary's role in reproduction leads to the development of the first hormone-based pregnancy test. Role of the corpus luteum is discovered. Estrogen and progesterone are isolated. Extra twenty-first chromosome (cause of Down syndrome) is discovered; method of counting chromosomes is established.	Patrick Steptoe performs in vitro fertilization. First intrauterine device (IUD) is developed for contraception. First contraceptive pill is introduced. Frozen spermatozoa are used for insemination. Amniocentesis is developed as a diagnostic tool. FDA approves the birth control pill. Giving birth in water gains in popularity. RU-486 is approved for use in France and China. The U.S Food and Drug Administration approves the use of contraceptive implants.	The first vaginal hysterectomy is performed with laparoscope. A breakthrough, nerve-sparing prostatectomy procedure is developed that helps avoid the impotence and incontinence associated with previous procedures. A new procedure for performing hysterectomy avoids incising the pelvic support structures, which helps avoid later complications such as vaginal, cervical, or uterine prolapse. Voice-controlled robots assist in a laparoscopic hysterectomy procedure.
2000–		The first male impotence drug is introduced on the commercial market. Reproductive cloning of humans is opposed worldwide. RU-486 is approved for use in the United States.	First total hysterectomy via laparoscope is performed: cervix, uterus, and ovaries are removed. Minimally invasive robotic surgery with cameras and computer technology is performed. Hysterectomy mortality rate is about .23 percent.

9

Reproductive Medicine in the Twenty-First Century

Many diseases of the human reproductive system are related to cancer or to sexually transmitted infections, and they, along with others arising from different causes, are examined in this chapter. Infertility is explored separately in Chapter 10. The information about prostate disorders presented here supplements more detailed discussions found in the Urinary System volume of this series.

This medical summary is based on widely available data from the National Institutes of Health, the American Cancer Society, the Centers for Disease Control and Prevention, the World Health Organization, and miscellaneous health-related sources.

CANCER

The National Cancer Institute describes cancer as a group of many diseases sharing a common trait: the overproliferation of cells. Sometimes cells can form a benign mass, in which the cells do not spread to other parts of the body. But other times, the normal genes that help regulate cell growth become damaged or mutant, perhaps because of a genetic predisposition or the effect of **carcinogenic** chemicals. These defective genes cause normal cells to reproduce without control, while other defective genes fail to suppress the unregulated growth. Proliferating faster than they can die, the cells form a malignancy, a mass known as a tumor, growth, neoplasm, nodule, or polyp. This mass, which can invade and destroy nearby tissues, is cancer. In **metastasis**, it spreads when malignant cells break away from an original tumor and are carried in the blood or lymph to distant body sites where they grow a secondary tumor. If a primary tumor of the breasts metastasizes to the colon, it is known as metastatic breast cancer—the abnormal cells in the colon are the same as the abnormal cells in the breast.

Because each cancer patient responds a little bit differently to different cancer therapies, **oncologists** use an assortment of weapons to fight the disease. Each case of cancer must be staged (see "Cancer Staging") to establish how far it has spread so that an appropriate treatment strategy can be developed. It will usually include **chemotherapeutic** drugs that kill both healthy cells and cancer cells. Like hormones or other substances introduced into the bloodstream, these chemicals are a form of systemic therapy that affect the whole body and frequently have significant side effects. Some of these include nausea and vomiting, hair loss, fatigue, and lowered immune resistance. Surgery, which is aimed directly at the cancer site and bypasses the rest of the body, is localized therapy, and radiation treatments can be either, depending on how and where they are administered. The side effects local treatments cause tend to be associated more closely with a specific site rather than the entire body. Newer cancer treatments, some of them in **clinical trials** and others of them already in use in the general population, include:

- monoclonal antibodies, molecules that attach themselves to cancer cells to invite attack from the patient's immune system;
- viral "smart bombs" that replicate in the body to overwhelm cancer cells;
- antiangiogenic drugs that kill tumors by blocking or destroying their blood supply;
- hormones that prevent or slow tumor growth; and
- nanocapsules or nanoshells, microscopic containers inserted into the body to deliver radioactive material directly to tumors.

Despite the promise of these treatments, many cancers still cannot be cured. Moreover, while survival rates are improving due in part to earlier detection, the **incidence** of some cancers is increasing. Although the quest for cures continues to drive research, a parallel focus of many experts is to emphasize prevention and to produce strategies that help patients manage existing disease. To this end, physicians urge their patients to make sensible lifestyle changes, especially quitting smoking, and to take advantage of modern screening technologies that detect cancers early.

Figure 9.1 reflects the estimated incidence and **mortality** rates for major reproductive cancers in the United States in 2003.

Cancer Site	Incidence	Mortality
Breast, Female	211,300	39,800
Cervix (Invasive)	12,500	4,500
Ovary	25,400	14,300
Prostate	220,900	28,900
Testicle	7,600	400
Uterus	40,100	6,800

Figure 9.1. Estimated annual incidence and mortality of major reproductive cancers, United States, 2003.
Based on data from the National Cancer Institute and the American Cancer Society.

Cancer Staging

How cancer is staged—how specialists define its extent—varies depending on the type of cancer. Certain types of leukemia may be staged by blood cell analyses, while tumors are staged by their size and the degree to which they have invaded normal tissue. The staging process is technical, and the breakdown here has been simplified for illustrative purposes. In most cases, the five levels are determined by three identifiers: the size of the tumor, the degree of lymph node involvement, and whether the cancer has metastasized. Specialists further refine their diagnoses with staging subcategories to help them treat the disease and predict its course.

Stage 0: A small tumor is found but it is contained deep within the tissue from which it arose.

Stage I: The tumor is small but has begun to spread into adjacent tissues.

Stage II: The tumor has exceeded a specific size and has invaded nearby tissue, but not the lymph nodes.

Stage III: The cancer has spread to the lymph nodes.

Stage IV: The cancer has spread to distant parts of the body.

Common Risk Factors for Women

Most cancers of the female reproductive system have several risk factors in common. Those that are unique to individual cancers are discussed in relation to specific organs. The most common factors are:

- increasing age;
- family history of the disease;

- genetic predisposition;
- onset of menstruation before age 12 or menopause after age 55;
- never having children (because women who have never borne children are exposed for a longer period to the estrogen their ovaries produce);
- use of hormone replacement therapy (HRT);
- history of radiation treatment to the area;
- race (Caucasian women are at highest risk for uterine, breast, and ovarian cancer, followed by African American women for uterine and breast and Hispanic women for ovarian cancer; Hispanic women are at highest risk for cervical cancer, followed by African American women);
- obesity associated with a high-fat diet;
- overindulgence in alcohol.

Breast Cancer

Cancer experts know that bumping or bruising the breasts does not cause breast cancer, but they do not know precisely what does. The National Cancer Institute points out that many women with multiple risk factors do not get breast cancer, while many women with only one risk factor, their gender, do. Women with defects in specific genes—BP1, BRCA1, and BRCA2—are at particularly high risk.

Experts recommend that women receive diagnostic mammograms to detect changes in their breasts and to learn to examine their own breasts on a monthly basis. Because there is seldom pain associated with early breast cancer, symptoms to which women should be alert include:

- a lump or other change in the breast, nipple, or underarm area;
- change in the appearance or size of the breast or nipple;
- discharge from the nipple.

In a new diagnostic test called *ductal lavage* that helps detect early breast cancer, a solution is rinsed through the breast's ducts to retrieve cells for microscopic examination. If abnormal cells are found, magnetic resonance imaging (MRI) or ultrasound aid in making a diagnosis. A **biopsy** may be necessary to establish whether the tumor is malignant and whether it is estrogen dependent—whether it relies on estrogen for continued growth. If it does, some oncologists recommend their patients have their ovaries removed to stop production of the hormone. The lymph nodes may also have to be biopsied to determine if the cancer has metastasized outside the breast.

If surgery is required, it may be preceded by chemotherapy or radiotherapy to shrink the tumor, and will probably be followed with chemotherapy to ensure all cancer cells have been killed. The surgical procedures chosen are based on the size and location of the tumor and may range from a simple

lumpectomy, which may take place at the same time as the biopsy, to a total mastectomy, in which the entire breast and other tissues are removed. In such cases, many women elect to have plastic surgery to rebuild the breast.

PAGET'S DISEASE OF THE NIPPLE

Not to be confused with Paget's disease of the bone, this malignant breast disease is characterized by lesions on the nipple and areola. It represents only 1 percent to 4 percent of all breast cancer cases. It usually appears as a reddened, irritated area, often accompanied by itching or a bloody discharge from the nipple.

Gynecological (Women's) Cancers

Gynecological cancers affect the uterus, ovaries, cervix, vagina, and vulva. Although most of these occur in women over age 50, their incidence among younger women has been increasing (see the later section, "Human Papillomavirus").

UTERINE CANCER

This is the leading gynecological cancer in women. Women who used HRT after menopause have a higher risk, especially if they did not take progestin along with estrogen. Since millions of menopausal women were prescribed these hormones over the last several years, there is concern that the incidence of uterine cancer will rise. Regardless of their cause, most uterine cancers originate in the endometrium as **adenocarcinomas**. A small percentage are **sarcomas**. In addition to those cited earlier, risk factors for uterine cancer include a history of either endometrial **hyperplasia** or colorectal cancer; the relationship between the latter and uterine malignancy is not yet understood.

The unusual vaginal bleeding or discharge associated with uterine cancer appears relatively early in the disease. Pelvic pain and pain during urination or intercourse are other symptoms. To rule out cancer, a physician is likely to conduct a pelvic exam and collect cervical cells to look for cellular abnormalities. The physician may also collect tissue from inside the uterus for biopsy. Transvaginal ultrasound, a painless procedure in which an ultrasound device is inserted into the vagina to reveal suspicious growths, is another valuable tool for diagnosing uterine cancer.

While radiation or hormone therapy may be used to treat the disease, most women undergo a hysterectomy to have the uterus, and usually the fallopian tubes and ovaries, removed. If there is reason to suspect the cancer will spread, the lymph nodes and cervix may be removed as well. Radiation therapy used to kill any remaining cancer cells may be in the form of vaginal implants that remain in place for several days and require the "radioactive" patient to stay in a hospital.

OVARIAN CANCER

Often called a "silent killer" because it is so difficult to detect in the early stages, ovarian cancer is the second most common gynecological malignancy. When detected in its advanced stages, it kills about three-quarters of patients within five years. There was a significant breakthrough in 2003 when researchers announced the development of a blood test that detects ovarian cancer very early. It holds great promise, but it must be tested in clinical trials with large groups of women to ensure it does not produce false positive results that lead to needless surgery. Transvaginal ultrasound can also help detect tumors in the ovaries.

The causes of ovarian cancer are not entirely understood, although it is clear that a family history of breast or colon cancer raises the risk, while taking contraceptive pills and breastfeeding may lower it, presumably because these reduce the number of times a woman ovulates. As might be expected, taking fertility drugs, which increases ovulation frequency, may increase risk.

When ovarian cancer is advanced, the symptoms may include abdominal discomfort such as pressure, cramping, or pain; diarrhea or constipation; frequent urination; abnormal vaginal bleeding; and loss of appetite or weight fluctuations. Since these symptoms may also suggest benign conditions, standard diagnostic tests are recommended: pelvic exam(s), Pap test, and imaging studies such as ultrasound and a computerized x-ray called a *CT scan*. Finally, a tissue biopsy may be necessary, performed in an operating room since the ovaries are usually removed immediately if cancer is present.

Because ovarian cancer is difficult to detect in its more treatable stages, many at-risk women, even if they are of childbearing age, elect to have a **prophylactic** oophorectomy, the preventive removal of healthy ovaries along with the fallopian tubes, uterus, and cervix before disease has a chance to strike. In patients with cancer, oophorectomy may include removal of the lymph nodes and other tissues, and chemotherapy or radiation therapy could be added to the treatment regimen. Later, a second surgery may be scheduled so the surgeon can examine the pelvic area for any remaining sign of disease.

INVASIVE CERVICAL CANCER

In industrialized nations, the Pap smear has prevented countless patients from developing invasive cervical cancer by catching precancerous changes early. In this procedure, the physician lightly scrapes cells from the cervix for microscopic examination. In colposcopy, the doctor can stain cervical cells to distinguish normal from abnormal ones, and may also scoop some cells from inside the cervix or remove a larger piece of tissue in a procedure called *conization*. If the cells are abnormal, indicating **dysplasia** or

cervical intraepithelial neoplasia 1 (CIN 1), the physician can remove the tissue with freezing, **cauterization**, or laser techniques. If there is a chance abnormal cells have entered the uterus, a D&C may be performed in which the lining of the uterus is scraped.

There appears to be a link between sexual activity and cervical cancer. Women who begin having sexual intercourse before age 18 or who have had sexual partners who engaged in sexual intercourse at a young age are at increased risk. Moreover, a larger number of partners increases risk. Interestingly, although cervical cancer is rare in women younger than age 35, it began to increase significantly in this group after 1960, a decade that marked the beginning of the sexual revolution when young women engaged in sex more frequently, often with multiple partners.

The human papillomavirus (HPV) is also related to the development of cervical cancer, although not all women with cervical cancer test positive for the presence of the virus nor do all patients with HPV develop cervical cancer (see the later section, "Human Papillomavirus").

As it does with most cancers, smoking raises the risk of developing cervical cancer, and experts know that women who were exposed in the womb to diethylstilbestrol (DES), a drug given to their mothers to prevent miscarriage, are at greater risk. Some researchers believe contraceptive hormones increase a woman's chance of developing the disease, but there is conflicting evidence challenging this belief. Patients whose immune systems are weakened by disease or by anti-rejection drugs seem more susceptible to developing precancerous lesions of the cervix.

In most cases, cervical cancer is treated surgically, followed by chemotherapy or radiation to ensure all cancer cells are killed.

LESS COMMON GYNECOLOGICAL CANCERS

Fallopian tube cancer, extremely rare, is thought to arise not from the Fallopian tubes but from metastatic ovarian cancer. Sarcomas, **melanomas**, adenocarcinomas, and **squamous cell carcinomas** are different kinds of vaginal cancers, but these malignancies are uncommon. Young women between 12 and 30 years of age, especially those exposed to DES, may develop a rare vaginal malignancy called *clear cell adenocarcinoma*. Laser surgery can be used to kill the cancer cells, but, in more advanced cases, surgery that removes a portion of the vagina may be required, followed by chemotherapy and radiation.

Over 90 percent of vulval cancers in younger women are strongly associated with human papillomavirus infection. Other types of vulval cancer are melanomas, adenocarcinomas, and sarcomas. Skin discoloration and warts, moles, or lumps appearing on the vulva are collectively known as "vulval intraepithelial neoplasia" (VIN), precancerous conditions often marked by intense itching. A visual exam at annual gynecological exams will help alert physicians to unusual changes in the color or texture of the vulva. Like cervical cancer cells, those of the vulva can be stained with a dye to help identify them.

During a woman's reproductive years, rare malignancies called *gestational trophoblastic cancers* can arise from early embryonic tissue that has not developed properly. A molar pregnancy is tissue that, rather than developing into an embryo, organizes into a mass of grapelike chorionic villi called an *hydatid mole*. Other gestational trophoblastic cancers can arise from uterine tissues and the site where the placenta attaches to the uterus.

Diagnoses of molar pregnancy and other masses may be based on a combination of factors: severe nausea and vomiting, vaginal bleeding, and the abnormal results of blood tests and ultrasound imaging. On the other hand, some gestational tumors arise from tissues left behind after normal childbirth, and in these cases they are difficult to detect until they produce abdominal swelling or other unmistakable symptoms.

Gestational masses are usually removed by suction dilation and curettage for women who intend to have children, or by hysterectomy for others. Chemotherapy, sometimes combined with radiation, may follow to kill remaining cancer cells. A patient receiving comprehensive treatment usually has a good prognosis.

Genitourinary (Male) Cancers

Cancer of the prostate, which primarily affects middle-aged and older men, represents the majority of genitourinary cancers and about 25 percent of all cancers in men. Testicular cancer, a disease usually diagnosed in men 15 to 35 years old, makes up less than 6 percent; penile or scrotal cancers comprise the rest.

PROSTATE CANCER

Every year, hundreds of thousands of men around the world are diagnosed with prostate cancer. Despite its increasing prevalence, the death toll from prostate cancer has dropped a significant 20 percent in the last decade, due principally to early detection. Much of the credit goes to screening tests that measure prostate-specific antigen (PSA) levels. PSA is a protein produced by the prostate that increases in the presence of cancer. A simple blood test reveals PSA levels and, combined with a digital rectal exam to check for enlargement of the gland, can suggest or rule out malignancy.

PSA screening is recommended annually for all men beginning at age 45 to 50, especially since early prostate cancer may be symptomless, or for younger men with certain risk factors. Despite its diagnostic value, PSA testing has come under criticism after evidence showed that it misses a majority of tumors. The reason, some experts believe, is not that the test is faulty but that its interpretation has been skewed by setting the threshold of significant numbers too high. On the other hand, some feel that a lower threshold may lead to overdiagnosing and overtreating, causing needless surgery and suffering. Still others argue that many prostate cancers are so slow growing that the risk of treating them is greater than the risk of leaving them alone, especially if the patient is elderly and likely to succumb to another disease before prostate cancer threatens his health. Complicating the issue are reports from the National Cancer Institute announcing that some men have a genetic alteration causing their prostates to produce high levels of PSA in the absence of cancer. For these reasons, it is unclear just what the ultimate recommendations for safe PSA levels will be.

Some of the risk factors for prostate cancer resemble those for breast cancer:

- genetic mutations;
- increasing age;
- family history of the disease;
- high-fat diet;
- race (African American men are more vulnerable, and tend to develop the disease earlier in life).

Any of the following symptoms may indicate the presence of prostate cancer and should be evaluated by a physician:

- frequent urge to urinate, especially at night;
- inability to urinate, or difficulty starting or stopping urination;
- weak or interrupted flow of urine;
- pain and burning upon urination;
- difficulty achieving an erection;
- painful ejaculation;
- appearance of blood in urine or semen;
- pain or stiffness in the lower back, hips, or upper thighs.

Because some of these symptoms may mimic urinary tract infections (see the Urinary System volume of this series) or benign prostate enlargement (see the later section, "Benign Prostatic Hypertrophy"), anyone experiencing them should promptly consult a physician who may perform a rectal exam, a urinanalysis, and a PSA screening. If a growth is present, imaging tests and a biopsy may be necessary to determine whether it is malignant and, if so, its stage.

Surgery to remove the prostate may be indicated, preceded or followed by radiation, hormone, or biological treatments, or **cryosurgery**. In some cases, an **orchiectomy** may be performed to halt the body's testosterone production, much as the ovaries of a woman with breast cancer may be removed to reduce the patient's estrogen production.

Surgical procedures to remove the prostate vary, but some have left men with lifelong impotence and urinary incontinence. Fortunately, the advent of an innovative nerve-sparing surgical technique in the 1980s has allowed many patients to avoid these consequences.

TESTICULAR CANCER

Nearly four times as common in white men as in black men, cancer of the testicles used to carry a grim prognosis, but new drugs and earlier detection have, as with many cancers, improved outcomes significantly. In fact, if it is treated early enough, testicular cancer is curable.

Risk factors for this disease include a history of cancer in the other testicle or in the family, and human immunodeficiency virus (HIV) infection. In particular, patients with an undescended testicle, even if it has been corrected surgically, are at greater risk (see the later section, "Undescended Testicles").

Physicians recommend that all men examine their testicles monthly to detect early tumors, which can be as small as a pea or as large as an egg. While such growths are often painless and symptomless, others are associated with tenderness, a sense of heaviness in the groin area, a bloody discharge from the penis, or even enlargement of the breasts due to hormone imbalances. If the physician thinks a lump is malignant, the testicle will probably be removed because the biopsy procedure alone could release cancer cells into the bloodstream. Although removal of one testicle does not impair the ability to achieve an erection, it may result in infertility because nerve pathways controlling ejaculation are disrupted. As in prostate surgery, however, there are nerve-sparing surgeries that can preserve this function in many cases.

PENILE AND SCROTAL CANCERS

Cancer of the penis, although rare in industrialized countries, occurs slightly more frequently in uncircumcised men. It is also seen more often in men who once received high-dose ultraviolet treatments for penile **psoriasis**. It is usually characterized by growths or sores on the penis, bleeding, or unusual discharge. Like any cancer treatment, the therapies aimed at penile cancer depend on the extent of the disease. In the cancer's earliest stages, chemotherapy, biological treatments, or radiation may destroy the malignancy, but most often these therapies are used as an adjunct to surgery. In some cases laser surgery is used, sometimes portions of the penis are surgically removed, and, in advanced cases, the entire organ must be amputated. The lymph nodes surrounding the penis may be excised as well. If the penis must be removed, the patient does not lose urinary control since the sphincter muscle that retains urine is left intact.

Cancer of the scrotum usually occurs in middle-aged and older men and is strongly associated with occupational exposure. The workplace connection was first made in the eighteenth century when young chimney sweeps developed the disease in unusual numbers. Later analyses showed that soot contains carcinogenic hydrocarbons, which are abundant in many workplaces today where posted warnings alert workers to their dangers. Scrotal neoplasms may begin as painless ulcers or lumps and progress to larger lesions, or they may manifest themselves as swollen glands in the groin. Like other cancers, scrotal malignancies respond well to treatment if they are caught early.

SEXUALLY TRANSMITTED DISEASES (STDs)

Among the most common of all infectious diseases, STDs, which used to be known as venereal diseases, are caused by pathogens, disease-producing micro-organisms. They are most often spread by sexual contact, but some can be passed from mother to child during childbirth or breastfeeding, through shared **intravenous** (IV) drug needles, or through contact with infected blood. Ranging from relatively mild to fatal, they tend to be most common among sexually active teenagers and young adults, especially those who have multiple sexual partners.

Many bacterial STDs are treated with antibiotics, but there is considerable concern in the medical community about resistance. Some bacteria have mutated, changing their genetic makeup in order to resist modern antibiotics; until new drugs are developed that kill these mutant strains, the bacteria pose a very serious threat. Nor are there cures for viral infections like herpes or AIDS, although newer drugs that relieve the symptoms and slow viral replication continue to be developed. Yeast and parasitic infections respond to several types of medications.

While they cannot completely prevent the spread of STDs, the right kind of condoms greatly reduces the risk of transmission. Those made of latex or polyurethane provide the most impenetrable barrier. The natural-membrane condoms (lambskin) are less safe, although they are effective contraceptives. Statistics show that risk factors for contracting an STD are:

- adolescence, principally because teens may not understand how to protect themselves or may engage in impulsive, high-risk sexual behavior;
- unprotected sex with multiple partners;
- history of STD infection;
- compromised immune system; and
- intravenous drug use.

Disease	Incidence	Prevalence
AIDS/HIV*	40,000	900,000
Chlamydia	3,000,000	2,000,000
Genital herpes	1,000,000	45,000,000
Gonorrhea	650,000	Not Available
Hepatitis B (HBV)	120,000	417,000
Human Papillomavirus (HPV)	5,500,000	20,000,000
Syphilis	70,000	Not Available
Trichomoniasis	5,000,000	Not Available

Figure 9.2. Annual incidence and prevalence of major STDs, United States, 2003.

*Of the new cases, about half occur in people under 25 years of age. Based on data from the National Institute of Allergy and Infectious Diseases, National Institutes of Health, and the Centers for Disease Control and Prevention.

Of the roughly 15 million new cases of STDs per year in the United States, at least one-fourth occur in teenagers. Females suffer the most severe consequences because some infections lead to disabling pelvic inflammatory disease, sterility, even cancer, and can greatly complicate pregnancy and endanger the life of a fetus. The chart shows the incidence and **prevalence** of major STDs in the United States as of 2003 (see Figure 9.2).

Acquired Immunodeficiency Syndrome (AIDS)

The evidence is irrefutable that HIV is responsible for AIDS, a disease of the immune system. It is spread by contact with bodily fluids during sexual activity or in healthcare environments, or by sharing IV needles. Despite the concerns of some that the virus can spread through air, water, or even mosquito bites, the Centers for Disease Control and Prevention states that there is no data to support these fears, nor are there any recorded cases of infection from contact with saliva, tears, or sweat. The virus attacks cells of the immune system, compromising its ability to fight either the virus or the **opportunistic infections** like pneumonia or cancer that overwhelm the body's defenses. In 2002, HIV/AIDS-associated illnesses killed about 15,000 people in the United States and about 3 million people worldwide.

Someone newly infected with HIV might not have symptoms for years, although a simple blood test can detect the virus within a few weeks of exposure. When symptoms emerge, they are often flulike and feature persistent fatigue, swollen lymph glands, nighttime fevers, diarrhea, and weight loss. These same symptoms may also be associated with infections secondary to HIV. There is no cure for AIDS, but the death rate in the United States has been reduced by about 70 percent since 1995 thanks to newer drugs that slow its course.

AIDS was identified in 1981 when it emerged in America's homosexual community. It has rapidly spread among heterosexuals, male and female. Of the 42 million people living with HIV throughout the world, 50 percent are women. About 25 percent of babies with HIV contract the virus from their infected mothers. Although the method of transmission is not clear, it was announced in 2003 that administering the HIV-fighting drug nevirapine to women in labor could reduce transmission by 41 percent.

Chlamydia

Chlamydia, an infection spread by sexual contact and childbirth, is the leading bacterial STD in United States. About 3 million people were infected with chlamydia in 2003, and the prevalence of those with the disease but still untreated (many of whom may not know they are infected) is 2 million. Although its symptoms are milder than those of gonorrhea (see the later section, "Gonorrhea"), it can be extremely serious for women if left untreated because it may lead to pelvic inflammatory disease, threaten pregnancy, or cause infertility. Chlamydia can also cause a condition known as lymphogranuloma venereum that is common in tropical areas of the world. A potentially malignant disease that first begins as a genital blister, it generally responds to drug therapy or can be surgically treated.

Genital Herpes

The herpes simplex virus, especially the strain known as HSV-2, is a highly contagious disease causing painful genital blisters and sores that is transmitted during sexual contact. The lesions usually appear on the penis, vulva, and cervix, then become dormant and disappear. They reemerge in response to stress, illness, or other situations in which the patient's immune system is weakened. Although the disease is most contagious while the sores are visible, some people contract herpes without developing lesions, so health specialists urge sexual partners to use condoms regularly. Genital herpes is a widespread, stubborn virus that afflicts about 45 million Americans, with about 1 million new cases expected this year. Although there is no cure, there are antiviral drugs that help relieve symptoms and prevent the virus from reproducing. In the meantime, researchers are aggressively seeking a vaccine for this troublesome disease. A pregnant woman with herpes may have to have her baby delivered by Caesarean section because herpes can complicate childbirth and endanger the infant's life.

Gonorrhea

In men, gonorrhea usually causes a discharge from the penis and burning or pain during urination, but up to 50 percent of women may have no symptoms at all. This makes it imperative that men with the disease alert their female partners. Untreated, gonorrhea can spread infection into other reproductive organs to cause infertility and scarring and, if the infection is allowed to spread throughout the body, lead to arthritis or heart disease. It has routinely been treated with antibiotics, but antibiotic resistance has made the use of more toxic agents necessary. Research to develop a vaccine has been underway for a number of years.

Hepatitis

Meaning "inflammation of the liver," hepatitis is a group of viral infections, some very serious. Hepatitis A is most often a food-borne illness that can also be spread by contaminated water or drugs. The patient usually recovers fully. Hepatitis B (HBV), the most common form of the disease and one that kills about 5,000 people annually in the United States, is most often associated with sexual contact. It can be treated, but it may leave patients with chronic and perhaps severe liver disease, even liver cancer, and makes them more susceptible to HIV. It is not known how well condoms help protect against its transmission. However, there is a vaccine, which the Centers for Disease Control and Prevention (CDC) refers to as the first "anticancer vaccine" because it prevents the liver cancer associated with the virus. The CDC recommends that all children be vaccinated, and many states require it before students may enter public school.

Hepatitis C (HCV) can lead to cirrhosis or liver cancer. It is spread primarily through blood transfusions, childbirth, or intravenous drug use, but it can be transmitted sexually if infected blood is present. There is no vaccine for hepatitis C.

Human Papillomavirus

The human papillomavirus is the most common sexually transmitted disease, often—but not always—causing genital warts (condylomata acuminata), especially in women. There are creams and solutions that can be used to remove them, or a physician may elect to freeze or burn them off, but the virus remains. It cannot be fully prevented by the use of condoms because it can be transferred by contact with any HPV-infected area of the body. It is strongly implicated in the development of cervical cancer. There is a vaccine being tested for HPV, but the National Cancer Institute has reported that it may not be effective during ovulation.

Syphilis

A scourge for centuries until penicillin was discovered in 1928, syphilis is a disease that can infect the neurological system and cause insanity and blindness years after transmission. Passed on to sexual partners by contact with a characteristic sore (chancre) on the genitals, neither person may know the lesion is there until chills, fever, and a body rash appearing a few weeks after exposure alert diagnosticians to its presence. Thanks to antibiotics, syphilis can successfully be treated, although any organs already damaged from its effects cannot be healed. After decades of reduction in incidence, the number of reported cases of syphilis in recent years has risen. This alarms public health officials who fear the disease may have developed some resistance to antibiotics.

Two disorders that resemble early-stage syphilis in appearance are chancroids and granuloma inguinale, both of which occur in tropical parts of the world and announce their presence with **papules** or pustules on the genitals that become painful and ulcerated. Chancroids do not imperil long-term health, but granuloma lesions have been associated with malignancy. Fortunately, both diseases respond well to drugs.

Trichomoniasis

A flagellate **protozoan** is responsible for this disease. It may be symptomless in men or result in some urethral irritation, but in women can produce significant vaginal discharge and genital inflammation. If unchecked, it may affect the ureters, prostate, and bladder. An antiprotozoal agent is the standard treatment for this disorder. Even if a man has no symptoms, he can unknowingly transmit the disease to future sex partners, and so must be treated along with the partner with whom he shared the disease.

Other STDs

In men, candidiasis, which results from infection with the yeastlike fungus *Candida*, causes itching and burning on the surface of the glans of the penis; in women, it produces vaginitis or vulvovaginitis, disorders that commonly feature irritation or itching along with discharge or odor. Once the offending organisms are identified, a number of different lotions, tablets, or suppositories can be used to eradicate them.

Also known as "crabs," *Pediculus pubis* refers to crab lice parasites that infest pubic hair. They cause intense itching and irritation as they feed on the host's blood, reproducing and transferring from person to person during sexual contact. Topical ointments and lotions are used to treat the infestation, which is often associated with poor hygiene. Crabs are not life threatening, but, like any parasite, they can carry infection and cause their hosts considerable distress and discomfort.

Salpingitis, an inflammation of the Fallopian tubes, is also called *pelvic inflammatory disease* (PID), although PID arises from other conditions as well (see the next section, "Other Female Reproductive Disorders"). The organisms causing infection are usually introduced through intercourse and rarely affect women during pregnancy or after menopause. Women with IUDs are more prone to the infection because the IUD serves as a platform from which pathogens can more easily invade the tubes. With pelvic pain as its primary symptom, salpingitis requires prompt, thorough treatment to prevent very serious, perhaps fatal, consequences.

OTHER FEMALE REPRODUCTIVE DISORDERS

Breast Syndromes

There are several benign conditions of the breast that affect women during their lifetimes. Imaging tests and fine-needle aspiration aid diagnoses, although some cases may require surgery to ascertain that breast growths or thickenings are not malignant. Some common breast disorders include

- fibrocystic disease—marked by lumps or fluid-filled cysts that tend to subside after menopause;
- chronic mastitis—inflammation usually caused by a bacterial infection;
- fat necrosis—benign lumps of tissue that form as the result of an impact injury, especially in obese women;
- fibroadenomas—harmless tumors that some physicians remove to avoid later malignancy;
- intraductal papillomas—growths in the ducts that cause a discharge from the nipple;
- sclerosing adenosis—an often painful condition due to calcified lumps growing in the lobules of the breast.

Dyspareunia

A word whose roots come from the Greek term meaning "painful mating," this condition is characterized in women by unusually uncomfortable or painful sexual intercourse. It is often due to the hormonal changes of menopause that lead to vaginal dryness and **atrophy**, but its presence may signal more serious conditions. It usually responds to treatment that addresses its underlying cause.

Endometriosis

This disease is characterized by endometrial cells growing outside the body of the uterus, particularly in the ovaries or throughout the pelvis. When the cells lining the uterus are shed in menstrual fluid each month, they exit the body through the vagina. In endometriosis, they have no exit, so they build up and cause inflammation and pain. In its milder forms, endometriosis can be treated with hormones; more severe cases may require either conservative surgery that preserves childbearing capability or hysterectomy.

Menstrual Difficulties

Menstrual problems may be due to a number of causes, some normal, such as temporary hormonal fluctuations, and some indicative of disease, such as the pain associated with endometriosis. The most common, many of which require medical attention, are:

- premenstrual syndrome (PMS)—ranging from mild bloating to abdominal pain, migraine-intensity headaches, and mood swings;
- amenorrhea—the absence of menstrual bleeding;
- menorrhagia—excessive bleeding;
- metrorrhagia—irregular menstrual cycles or bleeding due to causes other than normal menstruation;
- dysmenorrhea—painful menstruation;
- retrograde menstruation—the backup of menstrual blood into the **peritoneal cavity**;
- vicarious menstruation—bleeding from a site other than the uterus;
- painful ovulation;
- anovulation—the absence of ovulation (see Chapter 10); and
- bleeding between periods.

Ovarian Cysts

Cysts on the ovaries are usually fluid-filled structures that often disappear on their own. They may develop when the corpus luteum fails to dissolve completely after ovulation or they may be due to a hormone imbalance. They may cause pressure in the lower abdomen or unusual pain during menstruation. If their symptoms continue and they do not degenerate or respond to hormone therapy, the cysts and sometimes the entire ovary must be removed. While ovarian cysts are normally not dangerous, some can twist, causing severe pain and bleeding, and these may require prompt surgical removal. An inherited autosomal dominant condition known as polycystic ovary syndrome characterized by hormonal irregularities leads to enlarged ovaries, lack of menstruation, and, if untreated, infertility.

Pelvic Inflammatory Disease (PID)

PID refers to infections of the female pelvic organs that arise from a variety of causes: sexually related disease; injury from abortion, IUD insertion, or other invasive procedures; or **douching**. Although PID can be present with no symptoms, it is usually accompanied by abdominal pain or painful intercourse, menstrual irregularities, and possibly fever. Because a variety of different organisms can cause PID, broad-spectrum antibiotics may be used to eliminate the offending pathogens. The patient's sexual partner should be treated as well, to prevent reinfection and avoid the long-term disability and illness that can result from ongoing PID.

Prolapse

This condition occurs when the supporting ligaments that suspend the uterus and cervix lose their elasticity due to the strain of childbirth, obesity, uterine fibroids, or aging. The organs may drop down from their normal position into the vaginal canal. This condition can be corrected with a supportive brace, exercises, **prostheses**, or surgery.

Toxic Shock Syndrome

Related to a woman's use of tampons during her menstrual period, toxic shock syndrome is a medical emergency. During her period, a woman's vagina provides a damp, warm breeding ground for *Staphylococcus*, a bacteria that can reproduce rapidly in the right environment and produce sudden low blood pressure, vomiting, high fever, flulike symptoms, and severe shock that can lead to death. Prompt medical treatment is essential. Women can help reduce their risk by avoiding the exclusive use of high-absorbency tampons.

Uterine Fibroids

These benign, fibrous growths occur in about 20 percent of women over age 30. They are rarely life threatening, but they can produce an enlarged uterus, pain, and pressure on the bladder, leading to frequent urinary tract infections. They tend to occur more often in African American women and those who are childless or obese. When they are troublesome, the traditional treatment has been hysterectomy, but a new procedure called arterial embolization blocks blood flow to the fibroids and causes them to shrink. Fibroids can result in childbirth difficulties and excessive bleeding depending on their location. They tend to shrink on their own when a woman reaches menopause and her levels of estrogen decline.

Vaginismus

This unusual condition is characterized by an involuntary tightening of the pelvic muscles that prevents vaginal penetration. Women suffering from this condition are unable to use tampons, to have a pelvic exam, or to have intercourse. Its specific causes are not known, but is believed to be related to emotional stress and trauma, and must be treated with a combination of physical and emotional approaches that address its underlying causes. The possibility for complete recovery from the condition is uncertain, especially if the reasons for its development are not clear.

Vulvovaginitis

Vulval or vaginal inflammations are often characterized by intense itching, redness, and irritation of the affected tissues. They can be caused by STDs, the chemicals used in douche preparations, harsh soaps or detergents, tight-fitting noncotton underwear, and the use of certain antibiotics that promote an ideal environment in the vagina for yeast to proliferate. Easily treated, vulvovaginitis should not be ignored because it can lead to more serious disease or trigger urinary tract infections.

OTHER MALE REPRODUCTIVE DISORDERS

Benign Prostatic Hypertrophy (or Hyperplasia)

Otherwise known as BPH, this refers to the benign enlargement of the prostate that begins in some men after they have entered their 40s. Although it not a threat to health, especially in its early stages, it can interfere with normal urination and should be treated with medication or surgery to avoid the subsequent development of other disorders.

Gynecomastia

Breast enlargement sometimes occurs in adolescent boys in response to the hormone fluctuations of puberty, but it may also occur in middle-aged men whose hormones are once again irregular. Gynecomastia is usually not a cause for concern, but it should always be evaluated since it could signal the development of a serious liver disorder.

Impotence

Also known as erectile dysfunction, impotence is the inability to achieve an erection sufficient for satisfactory intercourse or, if erection can be maintained, the inability to ejaculate. It has many causes: psychological issues, physical defects, surgery, impaired neuromuscular function, drugs, or stress. There are medications that treat some forms of the disorder.

Inguinal Hernia

This refers to a bulge of tissue, usually intestinal, protruding through the muscles of the lower abdomen into the scrotum. Although benign, it can compress the sperm cord that houses nerves, blood vessels, and the vasa deferentia, so it should be surgically repaired.

Penile Lesions

If inflammations appear on the penis that are not caused by cancer or an STD, they are most likely due to yeast or bacterial accumulations beneath the foreskin of uncircumcised men. Antibacterials or antifungals are used to treat them before they can lead to more serious diseases.

Peyronie's Disease

This is a difficult-to-treat disorder seen in adult males. It is characterized by the collection of fibrous tissue in the penis that contracts, often causing the organ to curve into an arc. Although the thickened area is benign, it can be painful. Some cases may arise from physical trauma to the penis that causes internal bleeding. Others may be due to autoimmune disorders or the side effects of medications prescribed for certain conditions. In severe cases in which sexual intercourse is impossible, surgery may be necessary. Regardless of treatment, the outcome is often unpredictable.

Phimosis

This congenital or inflammatory condition causes the foreskin of the penis to tighten so it cannot be retracted over the glans. In paraphimosis, a retracted foreskin cannot be returned to its normal position. Circumcision is usually the recommended treatment.

Priapism

Unrelated to sexual arousal, priapism is a painful and prolonged erection due to the collection of blood in the erectile tissue of the penis. Its cause is unknown, although certain cerebrospinal conditions, infections, leukemia, trauma, or lesions within the penis are implicated in some cases. Temporary impotence is common after an episode of priapism, and healthy sexual function can be compromised unless the condition is treated promptly. Sometimes surgery is required to restore normal blood circulation to the penis.

Scrotal Growths

One of the most common types of benign scrotal masses is a hydrocele, a watery cyst in the scrotum. Others are hematoceles, containing blood, or spermatoceles, containing sperm. Fine-needle aspiration sometimes removes the cysts, but surgery is often required to prevent their return.

Undescended Testicles

Fetal testicles that do not drop into the scrotal sac shortly before the infant is born are known as undescended testicles. The problem often resolves itself but, if it does not, surgery can correct the condition. Adults with an undescended testicle have a higher risk of testicular cancer (see the earlier section, "Cancer").

Varicocele

A varicocele is a collection of enlarged blood vessels in the testicles, usually seen in men between the ages of 15 to 25. For unknown reasons, varicoceles tend to occur most often on the left side. They can be treated surgically if they produce pain or interfere with fertility.

SUMMARY

The medical conditions included in this chapter do not represent every disease or disorder of the human reproductive tract, but they introduce many of the most common. For more information, consult the relevant Web sites and other references at the end of this volume.

10

Infertility and Assisted Reproductive Technologies

SOME STATISTICS

Statistics from the Centers for Disease Control and Prevention (CDC) and similar organizations indicate that about one in ten couples has some problem with infertility. Not all cases must be treated with fertility drugs or assisted reproductive technology (ART) such as in vitro fertilization. Simple measures like carefully timing intercourse with ovulation may solve the problem for some. But for those who need additional help, there are several avenues to successful pregnancy. A child born of assisted reproductive technology today could have as many as five different parents instead of the traditional two:

- the biological mother—the woman who furnishes the egg;
- the biological father—the man who furnishes the sperm;
- the surrogate mother—the woman who gestates the embryo conceived in vitro from the egg of the biological mother;
- the nonbiological mother—the woman who raises the child;
- the nonbiological father—the man who raises the child.

Before the birth of Louise Brown in 1978, the first successful result of assisted reproduction, the most effective treatment for infertility was giving women hormones to stimulate their egg production, thus increasing the chance that sperm would encounter a viable ovum. The problem with this technique is that it often works too well: sperm fertilize several eggs, leading to multiple births. An advantage of the newer procedures, although they are more expensive, is that the number of embryos implanted in the mother can be regulated. Alternatively, excess embryos can be destroyed in the uterus if a woman carrying multiple fetuses elects to undergo a selective pregnancy reduction. Ideally carried out between nine and twelve weeks of pregnancy, it is an outpatient procedure in which potassium chloride is injected directly into the fetus, which destroys it. The risks to remaining fetuses are small, although the procedure increases the chance that the mother will go into premature labor.

ARTs, once considered experimental, now represent standard therapy for infertility. The CDC reports that the per-menstrual-cycle success rate for those who receive treatment is about 25 percent, rivaling that of fertile couples after one month of unprotected intercourse, with an overall success rate for couples completing three or more cycles of just over 50 percent. Of those pregnancies, about half result in multiple births.

In the spring of 2004, the President's Bioethics Advisory Commission reported concerns over the long-term health of children born of assisted reproductive technologies. In advising that studies be conducted to evaluate this issue, the commission also recommended that limits be imposed on how long leftover embryos may be stored for future fertilization attempts.

The individual statistics behind the success rates of ART vary to some degree based on:

- the cause of the infertility,
- types of drugs administered,
- the ART technique used,
- the patient's age,
- whether the eggs are donated, and
- whether the embryos were frozen before implantation.

Nearly 100,000 cases of infertility are reported in the United States every year. About three-quarters of the treatments use the patient's own eggs, fresh and newly fertilized, compared to about 13 percent that use the patient's thawed embryos. The rest use donor eggs.

At $6,000 to $10,000 per cycle for IVF treatments (see "Costs Associated with One Course of ART"), some insurance companies are reluctant to fund the procedures, defining them as "investigative" and "medically unnecessary" to avoid covering them. Because little more than half of the patients undergoing three or more cycles of ART realize a successful pregnancy, treatments represent a significant financial investment toward an uncertain reward. This can be a particularly large burden for young couples with lim-

Costs Associated with One Course of ART

A course of treatment generally lasts three to six consecutive menstrual cycles (months). Although costs vary widely throughout the world, fertility treatment in the United States for one cycle may range from $6,000 for fertility drugs alone to $10,000 for ART. Thus the potential costs for one course, after which specialists usually recommend the patient wait several months before another attempt, might range from $18,000 to $60,000. In some areas, it could be significantly higher.

ited resources. Nevertheless, the CDC reports that the number of couples seeking ART increases markedly every year.

This chapter discusses the most frequent causes of infertility and the tests that determine them, and introduces various ART procedures along with some of their accompanying risks and side effects.

WHAT IS INFERTILITY?

Strictly speaking, infertility—widely regarded as a disease among many medical professionals—is defined as an inability to conceive a child after one year of unprotected intercourse. Primary infertility is the inability to conceive a first child; secondary infertility is the inability to conceive again after one or more successful pregnancies. By comparison, sterility is generally understood to be a permanent condition, although certain sterilization procedures such as vasectomy can, in some cases, be reversed to restore an individual's fertility.

There is a common misconception that most causes of infertility reside with women. In fact, causes are evenly split—40 percent to 40 percent—between men and women. In the other 20 percent of cases, no cause can be found. Although couples under age 35 are considered infertile if they have not conceived within one year, couples over age 35 are usually advised to seek infertility counseling after six months, because their advancing age decreases their chances of successful conception and pregnancy.

Although there is no definitive measure of the impact psychological issues have on fertility, there is a widespread belief among specialists that stress, undue anxiety about the infertility itself, or joyless efforts to conceive can undermine a couple's attempts to do so. Reproductive specialists are generally equipped to address these problems or to recommend appropriate counseling.

CAUSES OF INFERTILITY IN FEMALES

Although there are many causes of female infertility, most cases can be traced either to ovulatory disoders or to blockages in the Fallopian tubes.

Ovulatory Disorders

Irregular or absent ovulation (anovulation) associated with infertility can often be treated with ovulation-inducing drugs, usually human menopausal gonadotropins (hMGs) that supplement or mimic the effects of follicle-stimulating hormones and prompt the ovaries to produce several eggs during a single month. Once the eggs have developed, human chorionic gonadotropin (hCG) is administered to trigger eggs' release into the Fallopian tubes where they can be fertilized.

There are a number of drugs marketed under different brand names that fertility experts use to direct the body's reproductive machinery. Many more are under development that promise both a greater degree of control over ovulatory events and a reduction in the often powerful side effects the drugs can cause (see the later section, "Risks and Side Effects of Treatment").

Anatomical Obstructions

Many cases of female infertility are attributed to obstructions in the Fallopian tubes or anatomical disorders of the uterus. Pelvic inflammatory disease and STDs are notorious culprits that scar and block the tubes, and sometimes these problems can be corrected surgically. Others, such as a double uterus (bicornate uterus), cannot. In some cases, a woman who has had a tubal ligation can have the operation reversed, but often this is not possible.

Cervical Receptivity

Sometimes the cervical mucus that protects the uterine cavity from the influx of bacteria also blocks the entrance of sperm. A simple midcycle analysis of the mucus should reveal whether this is a problem. If so, certain agents may be administered to thin the mucus and allow the passage of sperm, or alternative means of injecting the sperm into the uterus, such as intrauterine insemination (see the later section, "Diagnoses and Treatments"), may be used.

Age

After women enter their mid-30s, their fertility decreases; ovulation is less frequent and hormonal irregularities may upset the delicate balance required to sustain a successful pregnancy. In addition, the egg quality of older women degrades, increasing the likelihood that an egg successfully fertilized may contain abnormal chromosomes. Even if a pregnancy results, the embryo may abort.

Disease

Endometriosis has long been associated with infertility; recently one of the reasons has been discovered. A gene that normally produces an enzyme allowing the embryo to attach to the uterine wall is defective in women with endometriosis. Without that enzyme, the successfully conceived embryo is doomed. Certain other diseases, such as polycystic ovaries, may affect a woman's egg development, and conditions such as cancer of the reproductive organs, and its treatment, may interfere with her ability to engage in intercourse.

Other

There can be other causes of infertility, especially the temporary difficulty some couples face due to stress, poor diet, or excessive athletic activity in women that suppresses ovulation. Sometimes a pituitary gland tumor or other serious disorder of the endocrine system may be at fault.

CAUSES OF INFERTILITY IN MALES

Infertility in males is most often due to oligospermia or azoospermia (respectively, poor or absent sperm production) arising from the variety of factors discussed below. Other causes are related to impotence or an inability to ejaculate normally.

Hypogonadotropic hypogonadism is the failure of the testicles to produce sperm. Developing from a hormone deficiency that originates in the pituitary gland or the hypothalamus, the condition is sometimes treated with the same fertility drugs given to women, even though the U.S. Food and Drug Administration (FDA) has not approved them for this use. For unknown reasons, the drugs do not work as well for men, nor are they appropriate for every kind of hormone deficiency that affects male fertility. Other forms of hypogonadism may also decrease male **potency** and fertility. Some of these conditions are genetic in origin.

Genetics

Some estimates state that 10 percent to 25 percent of the cases of male infertility arise from deletions on the Y chromosome in overlapping areas known as the AZF (for the "azoospermia factor") regions, which have to do with sperm production. Other genetic causes are related to chromosomal abnormalities, which appear more often in azoospermic/oligospermic men. It is for this reason that many fertility clinics conduct genetic screening in such men to determine if there are mutations that could be passed on via fertilization.

Environmental Contaminants

Toxins such as lead and pesticides are notorious for impairing sperm count and **motility**. Because sperm takes three days to mature, many specialists recommend that men avoid exposure to any questionable materials for at least three days before submitting sperm for analysis or insemination. Men are also advised to ejaculate about three days before submitting a sperm sample so the sperm is fresh.

Drugs

It is not only drugs like those prescribed for high blood pressure that negatively affect fertility or sexual performance. Alcohol, marijuana, and nicotine also impair fertility, as do more potent drugs like cocaine or heroin. Because drug abuse affects the entire body, the extent of its impact on spermatogenesis will vary with each individual, but it can be significant.

Disease

Mumps, in particular, and any other disease that causes inflammation of the testicles can reduce sperm count or retard motility. STDs can cause scarring or other injuries, and diseases resulting in a high fever might disrupt spermatogenesis. Diabetes, cancer of the endocrine glands or reproductive organs, chemotherapy, and coronary artery disease also reduce sperm quality.

Anatomical Obstructions

A varicocele, a group of enlarged blood vessels that collects in the testicle, is one of the most common anatomic disorders that affects male fertility, presumably because impaired blood supply leads to the excessive heat that endangers sperm production. Tumors and hernias result in obstructed pathways, and undescended testicles surgically repositioned any later than early childhood will fail to produce sperm. A man who has had a vasectomy will be unable to conceive unless the procedure is reversed.

Ejaculatory Difficulties or Impotence

Sexual problems represent about 5 percent of infertility problems in men. Premature or retrograde ejaculation (see Chapter 2) are common ejaculatory problems, and impotence makes vaginal penetration impossible.

Other

Stress or a poor diet can affect sperm quality. Tight fitting clothing, especially underwear, can also be a factor.

DIAGNOSES AND TREATMENTS

Both women and men are advised to first consult their regular physicians if they are having difficulty conceiving. Sometimes there are simple explanations. Men who have a fever or are exposed to excessive heat in the scrotal area will have a significantly lowered sperm count, a temporary condition that will correct itself. In other cases, avoiding ejaculation for two or three days before ovulation to allow maximum sperm development and maturation may help. Indicators such as the woman's body temperature and the consistency of her cervical mucus can help couples pinpoint ovulation and establish an optimum time for intercourse. If these measures fail, they can be referred to appropriate fertility specialists who may include a reproductive endocrinologist, an embryologist, a gynecologist or obstetrician, and a urologist, each consulting with the patient(s) as the situation demands.

Women

The team will determine if an infertile woman is ovulating normally and, if not, a series of hormonal tests will be performed to determine the cause. Once the problem is corrected and normal ovulation returns, fertility is likely to be restored.

If this fails, the woman will receive gonadotropins (FSH and LH) to stimulate ovarian follicle growth and maturation, followed by another hormone to stimulate ovulation. Because these added drugs can cause significant side effects, her partner may also be examined to establish that the infertility does not originate with him.

Any woman taking fertility drugs must be monitored carefully, because the ovarian hyperstimulation they cause can be dangerous. If she conceives, she may be at risk for multiple births. If she does not conceive, the next step might be a hysterosalpingogram to rule out an obstruction in her Fallopian tubes.

In this procedure, the physician injects a dye to make the Fallopian tubes and other organs visible and to determine the site of any blockage. If further tubal or anatomical tests are required, a laparoscopic examination can be conducted during which the physician examines the organs for other problems. Finally, if there is reason to suspect a uterine disorder, an endometrial biopsy will be performed to check the uterine lining for abnormalities. Many can be surgically treated.

Until fairly recently, in vitro fertilization techniques followed the same pattern: after a woman's ovaries have been stimulated to produce several mature follicles, the patient is sedated or anesthetized. A needle inserted through her vagina extracts the eggs from the mature follicles. They are placed in a laboratory dish and mixed with sperm to be artificially inseminated. Once the fertilized eggs become several-celled embryos, the reproductive team determines through preimplantation genetic testing (see Chapter 7) which embryos are healthy and most likely to successfully implant. It is also at this stage that clinics allowing parents to specify the gender of their children will distinguish male from female embryos to select the ones to insert into the uterus.

Scientists have developed numerous refinements to early IVF technology:

- Intracytoplasmic sperm injection (ICSI) is used if the infertility originates with the male. His sperm is injected directly into the eggs within the laboratory dish.
- Transferring many-celled blastocysts from the lab dish to the uterus has, in many clinics, replaced an older method of transplanting eight-celled embryos, because blastocysts have better chances for survival.
- Assisted hatching is a microsurgical procedure in which the specialist dissolves a tiny part of the zona pellucida so the blastocyst has a better chance of implanting in the uterus.
- Gamete intrafallopian transfer (GIFT) allows physicians to promote fertilization by inserting the woman's eggs and her partner's sperm together into one of her Fallopian tubes.
- Zygote intrafallopian transfer (ZIFT), which is also called *tubal embryo transfer*, combines the technology of IVF and GIFT by inserting lab-fertilized eggs into the Fallopian tubes, rather than into the uterus.
- Intrauterine insemination (IUI) is a procedure usually attempted in couples whose infertility has no discernible cause. The woman is given fertility drugs to stimulate ovulation while her partner's sperm is injected directly into the uterus.
- Because major drawbacks to IVF are the risks and side effects of ovarian hyperstimulation, a new technique called in vitro maturation (IVM) has been developed that relies on natural cycles. Physicians remove soon-to-be-mature egg follicles prior to ovulation and allow them to ripen in the laboratory. Normally, they can retrieve about fifteen eggs, of which half may mature. Of those, five or six may become fertilized, and two may be implanted.
- A so-called "embryo glue" has been developed to help the blastocyst attach to the uterine wall.
- Sperm or egg donation as another route to pregnancy has become routine in cases in which the couple's gametes fail. This is often the case in older women whose eggs may have chromosomal damage. The in vitro fertilization and implantation procedures remain the same; only the source of the gametes changes.

Men

If a man's infertility is due to poor sperm production or quality, a semen analysis will be necessary. He collects a sperm sample through masturbation or intercourse using a special collection condom. The sperm is tested to ascertain its quantity, concentration, size, shape, and ability to move through cervical mucus.

Endocrinologic studies assessing thyroid, adrenal, and pituitary function are also likely, and the fertility specialists may wish to examine tissue from the testes. Except for instances of blockage, male infertility due to impaired sperm quality or quantity cannot always be treated directly, but the assisted fertilization techniques outlined above have helped to overcome this obstacle.

RISKS AND SIDE EFFECTS OF TREATMENT

The possibility of multiple births poses the biggest threat both to the mother and to the fetuses. Multiple-birth babies are often of lower birth weight, which predisposes them to developmental problems of varying severity, and the physical, psychological, and financial burdens placed on the mother—indeed, on the entire family—can be enormous. In Europe, where IVF may be covered by national health plans, fertility experts often implant only one embryo per cycle to eliminate the chance of multiple births. In the United States, where the number of cycles a couple can afford is limited, several embryos are usually inserted.

Whenever invasive procedures are performed on the human body, especially if anesthesia is required, there is risk. The systemic drugs used to adjust delicate hormone levels or stimulate egg or sperm production are very powerful, often having disturbing side effects for both men and women including breast swelling and tenderness, blurred vision, insomnia, nausea and vomiting, bloating, weight gain, mood swings, and, in women, ovarian cysts, although these are rare. As mentioned earlier, the additional danger of ovarian hyperstimulation is very serious.

Older women often seek assisted reproductive counseling because their eggs are no longer as viable. If they do become pregnant, the embryo is at a higher risk of being defective in some way. From age 30 to 35, a woman's chance of conceiving a child with Down syndrome jumps from 1 in 1,000 to 1 in 400; by the time she is 40, the risk has increased to 1 in 100, and to 1 in 30 when she is 45. A technique that has been developed to address an older woman's degraded eggs is cytoplasmic transfer, injecting the egg with cytoplasm from a young woman's egg. Presumably this confers more youthful properties to make the egg healthier and more fertile.

Cytoplasmic transfer also imparts the mitochondrial genetic material of the younger, third "parent" to the embryo. By means of a procedure involving nuclear transfer, U.S. and Chinese scientists reported in late 2003 that, for obstetrical reasons, they had to remove fertilized pronuclei from one ovum and insert them into another. Although the embryo in this procedure did not survive, it would have been born with genetic material from three parents. The latter procedure aroused controversy because it relied on a nuclear transfer technique that some find objectionable for ethical reasons.

Other problems associated with increasing age and pregnancy are high blood pressure, gestational diabetes, cardiac problems, and preterm labor. Nevertheless, many older women are now choosing to have children. This is due in part to women's longer lifespan and the greater economic independence current generations of women enjoy. Between 1990 and 1999, the number of births to women age 45 to 49 tripled. In many cases, they conceived through IVF techniques or they donated their eggs for implantation into a surrogate.

In recent years, it has even become possible for a postmenopausal woman to gestate an embryo and give birth if an ovum is donated, fertilized in vitro, and successfully implanted in her uterus. She must also receive appropriate hormones to support the pregnancy. It has also become possible, at least theoretically, for her to bear biological children. In an example of reproductive nuclear transfer, the nucleus of a menopausal woman's body cell would be transferred to another woman's enucleated (nucleus removed) healthy ovum. The egg would then be fertilized with sperm donated by the potential mother's partner. Transplanted into the self-donor or a surrogate for gestation, the embryo would be no different from its genetic sibling conceived during the couple's fertile youth. The FDA is evaluating this procedure.

Although emerging developments make it increasingly possible for postmenopausal women to have children, there are concerns about the wisdom of encouraging the practice for fear that older parents may not live long enough to raise young children to adulthood. This is a major reason many adoption agencies place age limits on potential adoptive parents.

OTHER ISSUES

Cultural, social, and legal issues complicate ART if it involves more than two "parents." Prebirth agreements and parental custody arrangements are frequently challenged in the courts. Judges and juries wading through the emotional and financial difficulties these cases present are tending to give greater weight to *intent* rather than *biology* in awarding custody; more and more, decisions generally favor the parent(s) whose clear intent is to rear and nurture the child, rather than the one who contributed genetic material or gestated the embryo. Nevertheless, definitions of parenthood are blurring, increasing the complexity of balancing parental "rights" against children's welfare.

SUMMARY

Infertile couples are able as never before to choose from an array of treatments to help them conceive. Administering hormones, engineering egg fertilization, and selectively harvesting tiny embryos for implantation have transformed human conception from a single sexual act into a finely calibrated series of steps in the laboratory.

Many regard the natural process of human conception and birth as a miracle. To some of those plagued by infertility, the mechanical interventions that make conception possible are no less miraculous.

Acronyms

AIDS Acquired immunodeficiency syndrome

ANS Autonomic nervous system

ART Assisted reproductive technology

AZF Azoospermia factor

BCE Before the common era

ß-hCG Beta human chorionic gonadotropin

BPH Benign prostatic hypertrophy (hyperplasia)

CAH Congenital adrenal hyperplasia

CDC Centers for Disease Control and Prevention

CE Common era

CIN Cervical intraepithelial neoplasia

CNM Certified nurse midwife

CRH Corticotropin-releasing hormone

D&C Dilation and curettage

D&E Dilation and evacuation

D&X Dilation and extraction

DES Diethylstilbestrol

DHEA Dihydroepiandrosterone

DHT Dihydrotestosterone

DNA Deoxyribonucleic acid

ELSI Ethical, Legal, and Social Implications Program

FDA Food and Drug Administration, U.S.

FSH Follicle-stimulating hormone

GIFT Gamete intrafallopian transfer

GnRH Gonadotropin-releasing hormone

HBV Hepatitis B virus

hCG Human chorionic gonadotropin

HCV Hepatitis C virus

HIV Human immunodeficiency virus

hMG Human menopausal gonadotropin

HMO Health maintenance organization

hPL Human placental lactogen

HPV Human papillomavirus

HRT Hormone replacement therapy

HSV Herpes simplex virus

ICSI Intracytoplasmic sperm injection

IUD Intrauterine device

IUI Intrauterine insemination

IUS Intrauterine system

IV Intravenous

IVF In vitro fertilization

IVM In vitro maturation

LH Luteinizing hormone

miRNA MicroRNA

MRI Magnetic resonance imaging

mRNA Messenger RNA

mtDNA Mitochondrial DNA

NVP Nausea and vomiting of pregnancy

PCR Polymerase chain reaction

PGD Preimplantation genetic diagnosis

PID Pelvic inflammatory disease

PMS Premenstrual syndrome

PSA Prostate-specific antigen

Rh Rhesus Factor

RNA Ribonucleic acid

RNAi Interference RNA

rRNA Ribosomal RNA

SHED Stem cells from human exfoliated deciduous teeth

STD Sexually transmitted disease

TENS Transcutaneous electrical nerve stimulation

tRNA Transfer RNA

VBAC Vaginal birth after Caesarean

VIN Vulval intraepithelial neoplasia

VIP Vasoactive intestinal peptide

VNO Vomeronasal organ

ZIFT Zygote intrafallopian transfer

Glossary

Adaptation Integral to the concept of natural selection, the tendency of an organism to change to suit its environment and increase its chances of survival.

Adenocarcinoma A cancer that starts in glandular tissue, like the ducts or lobules of the breast.

Afferent A term applied to the act of moving inward or toward a center; refers primarily to vessels and nerves.

Alleles Genes on homologous chromosomes that encode for the same trait. The genes for eye color—one on the maternal chromosome and one on the paternal—are alleles.

Amino acids The subunits or "building blocks" of proteins.

Analog An organ that is similar to another in function.

Antigen A foreign particle (e.g., a virus or bacterium) that causes an immune system response.

Aphrodisiac A substance used to heighten sexual desire.

Artificial intelligence The capacity of a machine to perform tasks traditionally thought to require intelligent awareness.

Asexual reproduction Reproduction in which genetically identical offspring are produced from a single parent.

Atrophy The wasting of a tissue, organ, or body part.

Biomarker A molecular clue indicating the presence of disease or the genetic predisposition for disease.

Biopsy The removal of a tissue sample from the body in order to examine it for disease.

Blastocyst An embryonic stage following the morula stage characterized by outer trophoblast cells, an inner cell mass, and a central, fluid-filled cavity.

Blastoderm The cell layer forming the wall of a blastocyst.

Carcinogenic A term describing any substance that can cause cancer.

Castration anxiety Specifically, a fear of the loss of or injury to the genitals. More generally, it refers to anxiety disorders related to the loss of personal safety or power.

Cauterization A medical procedure in which diseased tissue is burned away.

Centromere A specialized, constricted region on each chromosome; in cell division, sister chromatids are attached at the centromere.

Cervical intraepithelial neoplasia 1 (CIN 1) Abnormal cells suggestive of cancer of the cervix.

Chemotherapy The use of chemicals to treat cancer.

Chromatin Diffuse mixture of DNA and proteins that condenses into chromosomes prior to cell division.

Clinical trials Rigorously controlled studies in which experimental drugs and other therapeutic agents are tested on patients before they are approved for use by the general public.

Codon A group of nucleotide "letters" that "spell," or comprise, an amino acid.

Collective unconscious The term used to describe the ancestral memory of a given culture or species, composed of repressed or unconscious material not perceived on a personal level.

Colostrum Nutritious fluids secreted by the breasts shortly before and after a woman gives birth; precedes the production of breast milk.

Complementary base pair Nucleotide bases (adenine and thymine or guanine and cytosine) that pair up via hydrogen bonds in DNA.

Congenital An adjective describing a condition present at birth.

Cryosurgery A surgical procedure in which diseased tissue is frozen or subjected to cold temperatures. May be used to destroy disease or help heal damaged tissue.

Curette (or curet) A spoon-shaped surgical tool used for scraping tissue.

Cytoplasm The cellular fluid inside the membrane of the cell and outside its nucleus. It contains numerous organelles that carry out cellular functions.

Daughter cells Cells arising from mitotic division that are identical to the parent cell.

Diffusion A chemical process by which particulate matter in fluid moves from one area of concentration to another.

Dilation A widening, as when the cervix dilates to allow passage of the baby during childbirth.

Douching Rinsing the vagina with medicated or cleansing solutions; often prescribed to treat mild vaginal irritations.

Down syndrome Mental retardation associated with specific chromosomal abnormalities.

Dysplasia The abnormal development of tissue.

Effacement The thinning and shortening of the cervix that occurs during childbirth.

Embryogenesis The entire process of cell division and differentiation leading to the formation of an embryo.

Endocrine system The body's network of ductless glands that secretes hormones and other chemicals directly into the bloodstream. *See also* **exocrine system.**

Endoscopic Refers to an endoscope, a lighted instrument used to look inside the body, especially in hollow organs or bodily canals.

Epiblast The outer layer of a blastocyst before differentiation into the ectoderm, mesoderm, or endoderm.

Epidural The injection of local anesthesia into the epidural space of the spinal cord to numb a large portion of the body, as in childbirth.

Epigenetic code A regulatory system by which histones and other proteins affect gene expression.

Episiotomy An incision in a woman's perineum during childbirth that helps allow the baby to emerge from the birth canal without tearing the mother's tissues.

Epithelial Adjectival form of *epithelium*, the cells that cover the internal and external organs of the body and the linings of vessels.

Erectile dysfunction The inability to achieve or maintain an erection.

Exocrine system The body's network of glands that secretes hormones and other chemicals directly into tissue through ducts. *See also* **endocrine system.**

Expressed In genetics, a term describing the results of activating of a gene. For example, in someone actively suffering from Huntington's chorea, the gene for the disease has been expressed.

Fetal alcohol syndrome A general term describing a series of birth defects, which may include mental retardation or behavioral problems, that are due to maternal alcohol use or abuse during pregnancy.

Gametes Reproductive cells that, before fusing at fertilization, are haploid—they contain twenty-three instead of forty-six chromosomes.

Genetic drift Random changes in the frequency of alleles from generation to generation. In small populations, it may eliminate a particular allele.

Genetic imprinting Refers to differences in the way maternal or paternal genes are expressed in the offspring.

Genetic sex Gender determination based on an XX or an XY chromosome configuration.

Genome The entire set of genes found in the DNA of a particular species, such as those in the human genome.

Genotype The genetic (allelic) makeup of an organism with regard to an observed trait, such as eye color.

Glucose tolerance test A test measuring blood sugar levels that is often used to diagnose diabetes.

Graft rejection The tendency of the immune system to reject transplanted tissue as foreign.

Histones Proteins associated with gene expression.

Homeostasis The body's tendency to maintain a relatively constant internal environment.

Homologous In genetics, chromosomes (one from the male parent, one from the female parent) carrying alleles for similar traits, such as eye color, that pair up during meiosis.

Huntington's chorea A progressive and fatal disease affecting the nervous system.

Hyperplasia The overgrowth of cells or tissue in a specific area; although the cells are not abnormal, their presence may be an early warning sign of cancer.

Hypoblast The inner layer of tissue in a developing embryo that will eventually become the digestive tract and respiratory tract.

Iatrogenic Defines a disorder that was caused by medical treatment or exposure to a healthcare environment.

Immortal In biology, refers to the capability of cells to divide indefinitely.

In vitro Occurring outside the body, often used to refer to laboratory procedures such as fertilization of ova within a laboratory dish.

In vivo Occurring inside the body.

Incidence In determining the extent of disease, the number of new cases of a disease that occurs in a population in a given time period.

Incontinent Partial or complete inability to contain urine.

Inorganic Not related to or arising from living matter.

Intravenous (IV) Within a vein of the body.

Karyotype A depiction of the chromosomes in a cell, sometimes aligned in pairs.

Libido The collection of urges comprising sexual desire; the sex drive.

Lumpectomy Surgery to remove a breast tumor and a minimal amount of tissue surrounding it.

Mastectomy Surgery to remove all or part of the breast. Depending on the extent of disease, other tissue such as the nipple or lymph nodes may be removed.

Melanoma A malignant tumor in melanocytes, specialized skin cells that give skin its color.

Metastasis The spread of cancer cells to distant areas of the body by way of the lymphatic system or bloodstream.

Microarray analyses Technology that evaluates the activity and role of individual genes.

Midwifery Refers to the profession of those specially trained to assist women in childbirth.

Monozygotic Refers to twins arising from one ovum.

Mortality The death rate from disease within a given population.

Morula A compacted group of embryonic cells at a level of development between the zygote and blastocyst stages.

Motility A capacity for movement; measured in fertility testing to help determine the vitality of sperm.

Multiorgasmic Capable of having more than one orgasm in a short period of time.

Multiple marker test Testing to screen for various biomarkers of disease. *See also* **biomarker**.

Multipotent cells Adult stem cells that can give rise to certain types of other cells. *See also* **pluripotent cells; totipotent cells**.

Nanotechnologies Different technologies based on the principles or use of tiny particles.

Natural selection The process by which the more fit organisms in a group survive and reproduce, passing on to succeeding generations the very qualities that rendered them more fit and ultimately reducing the number of individuals who do not possess those qualities.

Neonatal Generally refers to the first few weeks of newborn life or, more specifically, to the first twenty-eight days after birth.

Neurohormones Hormones secreted within the nervous system, often to trigger the release of other hormones.

Neuron Specialized cells that transmit the body's messages throughout the nervous system.

Oncologist A doctor who specializes in the diagnosis and treatment of cancer.

Oocytes Ova that have not yet matured in the ovary; they arise from primordial oogonia that develop in the fetus.

Oogonia Cells that arise from primordial germ cells and differentiate into oocytes in the ovary.

Opportunistic infections Infections that develop because a weakened immune system, often under assault from another disease such as AIDS, is unable to mount a defense.

Orchiectomy Surgery to remove one or both testicles.

Organelle Living particles in the cytoplasm of a cell such as mitochondria or ribosomes.

Osmosis A chemical process by which a solvent passes through a semipermeable membrane to redistribute the concentration of a dissolved substance.

Osteoporosis A disease most commonly seen in postmenopausal women in which the bone loses minerals, making it weak and subject to fracture.

Ovariotomy Surgery in which the surgeon cuts into the ovaries. *Ovariectomy* or *oophorectomy* is the removal of one or both ovaries.

Papules A solid area of skin that can be felt above the surrounding skin and is usually small in diameter.

Parthenogenesis A form of asexual reproduction in which offspring are produced by an unfertilized female.

Pathogen A microbe or other organism that causes disease.

Penis envy A term in psychology that usually refers to female envy of male power, which is often symbolized by the penis.

Perimenopause A period of physiological and psychological changes in females when their production of sex hormones begins to decrease with age; the period before menopause.

Peritoneal cavity The abdominal area housing internal organs.

Phosphate A chemical related to energy usage and transmission of genetic information in the cell.

Pica A craving for substances other than food such as dirt, starch, or glue.

Pluripotent cells Embryonic stem cells that are found in a blastocyst. They can give rise to every tissue of the body except certain extraembryonic tissues and the placenta. *See also* **multipotent cells**; **totipotent cells**.

Polymerase chain reaction (PCR) Applied in technology that was invented to copy DNA rapidly in the laboratory, it is the tendency of one newly copied strand of DNA to provide the template for the next.

Polyspermy The entrance of several sperm into an ovum.

Posthumanism The concept that humans could be fundamentally changed by biotechnology that replaces some of their genes or organs with artificial or other nonhuman parts.

Postmortem After death.

Postnatal After birth.

Potency In sexual reproduction, the ability of a male to engage in intercourse and fertilize an ovum.

Prenatal Before birth.

Prevalence In determining the extent of disease, a measure of the proportion of persons in the population with a certain disease at a given time.

Prolapse The sinking or falling of an organ from its original site.

Pronuclei The male and female gametes prior to fusing with one another at fertilization.

Prophylactic Preventive.

Prostheses Artificial replacements for missing body parts, such as replacement joints.

Protozoan A single-celled microorganism.

Psoriasis A common skin disorder characterized by an overgrowth of epithelial cells.

Pubescent At or reaching the stage of puberty.

Puerperal fever A bacterial infection usually spread by unsanitary obstetrical conditions. In the nineteenth century, when hospital births became more common, it killed over 20 percent of mothers. Also called *childbed fever.*

Punnett square A diagrammatic method of determining the distribution of inherited traits in offspring given the genotypes of the parents. *See also* **genotype**.

Reprogenetics A modern term coined to refer to the combined science or study of reproduction and genetics.

Ribosomes Cellular organelles that aid in the production of proteins.

Sarcoma A malignant tumor growing from connective tissues, such as cartilage, fat, muscle, or bone.

Secondary sexual characteristics Physical characteristics, like breast development and the growth of pubic hair, that arise from hormonal changes associated with sexual maturation.

Sertoli cells Cells in the testicles that nourish early sperm cells.

Sex-linked inherited characteristics Traits, such as color-blindness, that are linked to genes on the sex chromosomes, especially the X chromosome.

Sickle cell anemia A serious autosomal recessive disease characterized by abnormal red blood cells.

Spermatogonia Primordial sperm cells that develop in the male fetus.

Squamous cell carcinoma A slow-growing cancer beginning in nonglandular cells of the body, like the skin or lungs.

Synthesize To combine or manufacture a chemical or product.

Topical In medicine, an isolated place on the body.

Totipotent cells The cells resulting from fertilization before the zygote becomes a blastocyst. These cells can create any tissue in the body. *See also* **multipotent cells** and **pluripotent cells**.

Toxoplasmosis A protozoal disease dangerous to the fetus if the mother has an acute infection; it is transmitted by parasites.

Transcribe The manufacture of an RNA copy of a gene; the first step in gene expression.

Transcription factor A protein or other molecule that controls transcription.

Transcutaneous electrical nerve stimulation (TENS) A method of pain relief delivered by electrically stimulating nerve endings under the skin.

Transgenic A term describing an organism having the genes of another species inserted into its genome.

Translate The process by which the genetic information carried by messenger RNA directs the manufacture of proteins.

Ultrasound scan An imaging method using high-frequency sound waves to form images inside the body. Also called *ultrasonography*.

Ureters The tubes leading from the kidneys to the bladder.

Vertebrates Animals such as reptiles and mammals that have a segmented spinal column.

Vestibule The opening or entrance to a passage, such as the vestibule of the vagina.

Vestigial A term for nonfunctional remnants of organs.

Zona pellucida The outer covering of an ovum.

Zygote A diploid cell resulting from fertilization of an egg by a sperm cell.

Organizations and Web Sites

Access Excellence @ the National Health Museum
www.accessexcellence.org/AB/BA/Gene_Therapy_Overview.html

This site provides an overview of gene therapy and a discussion of diseases arising from genetic origins.

Biology Online
www.biology-online.org

This site leads readers to hundreds of topics of general or specific interest via a biology dictionary, Web links, and tutorials. It is an excellent link to research sources in specific subject areas.

Chromosome Deletion Outreach, Inc.
P.O. Box 724
Boca Raton, FL 33429-0724
Phone: (561) 395-4252 (Family Helpline)
www.chromodisorder.org/intro.htm

A non-profit organization offering information on chromosome disorders including deletions, trisomies, inversions, and translocations.

Estrella Mountain Community College
Maricopa Community Colleges
3000 North Dysart Road
Avondale, AZ 85323
Phone: (623) 935-8000
www.emc.maricopa.edu/faculty/farabee/BIOBK/BioBookTOC.html

This on-line biology book offers clear, straightforward text and colorful graphics to explain difficult concepts and illustrate reproductive anatomy.

National Human Genome Research Institute
National Institutes of Health
Building 31, Room 4B09
31 Center Drive, MSC 2152
9000 Rockville Pike
Bethesda, MD 20892-2152
Phone: (301) 402-0911
Fax: (301) 402-2218
www.nhgri.nih.gov

The organization that led the Human Genome Project for the National Institutes of Health continues genomic research aimed at improving human health and fighting disease.

National Institute of Child Health and Human Development (NICHD)
P.O. Box 3006
Rockville, MD 20847
Phone: (800) 370-2943
Fax: (301) 496-7101
Email: NICHDInformationResourceCenter@mail.nih.gov
www.nichd.nih.gov

One of the National Institutes of Health, NICHD seeks to ensure reproductive health for women and developmental health for all children to ensure a productive life free of disease and disability.

Society for Developmental Biology
9650 Rockville Pike
Bethesda, MD 20814-3998
Phone: (301) 571-0647
Fax: (301) 571-5704
Email: ichow@faseb.org
http://sdb.bio.purdue.edu

Valuable to students interested in university-level research, this site links readers to information about advances in specific areas of biology.

University of California
College of Letters and Science
201 Campbell Hall
Berkeley, CA 94720-2920
Phone: (510) 642-4487
Email: jforte@uclink4.berkeley.edu
http://biology.berkeley.edu/bio1a/topic/Sex_Reproduction

Slides and movies of male and female reproductive anatomy and function are available on this visually appealing site.

University of Manitoba
Web Centre for Women's Health
Winnipeg, MB
Canada R3T 2N2
Phone: (204) 474-8880
Email: whealth@cc.umanitoba.ca
www.umanitoba.ca/womens_health

A Web site devoted to information about obstetrics, gynecology, and reproductive services.

University of New South Wales Embryology
Cell Biology Lab
School of Medical Sciences (Anatomy)
The University of New South Wales
Sydney NSW 2052
Australia
Phone: +61 (2) 9385 2477
Fax: +61 (2) 9313 6252
Email: m.hill@unsw.edu.au
http://anatomy.med.unsw.edu.au/cbl/embryo/Embryo.htm

Offering step-by-step information on the normal and abnormal development of human embryos (and certain other mammals), this site also discusses child development and women's and men's health issues. The sitemap is most useful for guiding readers to specific areas of development.

University of Plymouth
Department of Psychology
Drake Circus
Plymouth PL4 8AA
Devon, UK
Email: pkenyon@plymouth.ac.uk
http://salmon.psy.plym.ac.uk/year2/Sexbehav.htm

Offering a bibliography of sources for exploring the subject further, this site features a comprehensive analysis of the nature vs. nurture debate over sexual orientation and the hormonal factors involved in psychosexual development.

University of Toronto at Mississauga
3359 Mississauga Road North
Mississauga, ON, L5L 1C6
Email: webadmin@utm.utoronto.ca
www.erin.utoronto.ca/~w3bio380

An on-line biology course devoted to human development, this site offers clear graphics and text. Clicking on the "Lecture Schedule" offers students specific areas of interest to explore in detail.

U.S. Human Genome Project
Oak Ridge National Laboratory (ORNL)
1060 Commerce Park, MS 6480
Oak Ridge, TN 37830
Phone: (865) 576-6669
Fax: (865) 574-9888
Email: genome@science.doe.gov
www.ornl.gov/TechResources/Human_Genome/home.html

The home page of the Human Genome Project, this site offers a wealth of information about the biology and medical implications of gene technology as well as the ethical and social challenges it represents.

The Visible Embryo
www.visembryo.com

This informative site features a graphic spiral depicting an embryo during twenty-three stages of development. Comprehensive explanatory text accompanies the graphics.

Bibliography

Bagemihl, Bruce. *Biological Exuberance: Animal Homosexuality and Natural Diversity.* New York: St. Martin's Press, 1999.

Bainbridge, David. *Making Babies: The Science of Pregnancy.* Cambridge, MA: Harvard University Press, 2001.

Darwin, Charles. *The Origin of Species.* New York: P.F. Collier & Son, 1909.

Dawkins, Richard. *The Blind Watchmaker.* New York: W.W. Norton & Company, Inc., 1986.

———. *The Selfish Gene.* Oxford: Oxford University Press, 1989.

Dennett, Daniel C. *Darwin's Dangerous Idea: Evolution and the Meanings of Life.* New York: Simon & Schuster, 1995.

Ellison, Peter. *On Fertile Ground.* Cambridge, MA: Harvard University Press, 2001.

Fukuyama, Francis. *Our Posthuman Future: Consequences of the Biotechnology Revolution.* New York: Farrar, Straus & Giroux, 2002.

Futuyma, Douglas J. *Science on Trial: The Case for Evolution.* Sunderland, MA: Sinauer Associates, Inc., 1995.

Gorski, Roger A. "Development of the Cerebral Cortex: XV. Sexual Differentiation of the Central Nervous System." *Journal of the American Academy of Child and Adolescent Psychiatry* 37, no. 12 (1998): 1337–1339.

Gould, Stephen Jay. *The Structure of Evolutionary Theory.* Cambridge, MA: Harvard University Press, 2002.

Gray, Henry. *Anatomy of the Human Body,* 20th ed. Edited by Warren Harmon Lewis. Philadelphia: Lea & Febiger, 1918.

Hollen, Kathryn. "The Changing Focus on Cancer: Emphasizing Prevention vs. Therapy." *Primary Care and Cancer* 21, no. 3 (March 2001): 36–38.

Isaacson, Tiffaney, comp. "What the Numbers Say: Birth, Breastfeeding, and Family in America." *Mothering* (March-April 2002): 37–45.

Jacquart, Danielle, and Claude Thomasset. *Sexuality and Medicine in the Middle Ages.* Trans. by Matthew Adamson. Princeton, NJ: Princeton University Press, 1985.

Kimura, Doreen. "Sex Differences in the Brain." *Scientific American Special Edition* (June 2002): 32–37.

Krebs, Robert E. *Scientific Laws, Principles, and Theories: A Reference Guide.* Westport, CT: Greenwood Press, 2001.

Lyons, Albert S., and R. Joseph Petrucelli. *Medicine: An Illustrated History.* New York: Harry N. Abrams, Inc., 1987.

Margulis, Lynn, and Dorion Sagan. *Acquiring Genomes: A Theory of the Origins of Species.* New York: Basic Books, 2002.

Mayr, Ernst. *What Evolution Is.* New York: Basic Books, 2001.

McKibben, Bill. *Enough: Staying Human in an Engineered Age.* New York: Times Books, 2003.

National Cancer Institute. *What You Need to Know about Breast Cancer.* Bethesda, MD: National Institutes of Health, April 2003.

Pfaff, Donald. *Drive: Neurobiological and Molecular Mechanisms of Sexual Motivation.* Cambridge, MA: MIT Press, 1999.

Porter, Roy. *The Greatest Benefit to Mankind: A Medical History of Humanity.* New York: W.W. Norton & Company, Inc., 1997.

———, ed. *Medicine: A History of Healing—Ancient Traditions to Modern Practices.* New York: Marlowe & Company, 1997.

Rako, Susan. *The Hormone of Desire: The Truth About Sexuality, Menopause, and Testosterone.* New York: Crown Publishers, Three Rivers Press, 1996.

Ridley, Matt. *Genome: The Autobiography of a Species in Twenty-three Chapters.* New York: HarperCollins, 1999.

———. *The Red Queen: Sex and the Evolution of Human Nature.* New York: Penguin Books, 1993.

Rodgers, Joann Ellison. *Sex: A Natural History.* New York: Henry Holt and Company, 2002.

Stock, Gregory. *Redesigning Humans: Our Inevitable Genetic Future.* New York: Houghton Mifflin Company, 2002.

Stoppard, Miriam. *Conception, Pregnancy and Birth.* New York: Dorling Kindersley, 2000.

Strickberger, Monroe W. *Evolution*, 2nd ed. Boston: Jones and Bartlett Publishers International, 1996.

Sykes, Bryan. *The Seven Daughters of Eve.* New York: W. W. Norton & Company, 2001.

Weinberg, Robert. *One Renegade Cell: How Cancer Begins.* New York: Basic Books, 1998.

Windelspecht, Michael. *Groundbreaking Scientific Experiments, Inventions, and Discoveries of the Seventeenth Century.* Westport, CT: Greenwood Press, 2002.

Index

About the Author

KATHRYN H. HOLLEN is a freelance science writer and editor who works with her husband, a technical illustrator, out of her rural home west of Washington, DC. She writes for the National Cancer Institute and other organizations engaged in biomedical reporting and research.